PRACTICING GRAMMAR AND USAGE

Lois G. Reynolds
Pellissippi State Technical Community College

Prentice Hall Reference Guide

SIXTH EDITION

Muriel Harris

Upper Saddle River, New Jersey 07458

© 2006 by PEARSON EDUCATION, INC.
Upper Saddle River, New Jersey 07458

All rights reserved

10 9 8 7 6 5 4 3 2 1

ISBN 0-13-168036-6

Printed in the United States of America

This work is protected by United States copyright laws and is provided *solely for the use of instructors* in teaching their courses and assessing student learning. Dissemination or sale of any part of this work *(including on the World Wide Web)* will destroy the integrity of the work and is not permitted. The work and materials from it should never be made available to students except by instructors using the accompanying text in their classes. All recipients of this work are expected to abide by these restrictions and to honor the intended pedagogical purposes and the needs of other instructors who rely on these materials.

CONTENTS

Part One: The Writing Process
1. Purposes and Audiences — 1
2. Writing Processes and Strategies — 10
3. Paragraphs — 19
4. Argument — 28
5. Visual Argument — 32

Part Two: Sentence Accuracy, Clarity, and Variety
6. Comma Splices and Fused Sentences — 34
7. Subject-Verb Agreement — 35
8. Sentence Fragments — 38
9. Dangling and Misplaced Modifiers — 40
10. Parallel Construction — 43
11. Consistency/Avoiding Shifts — 44
12. Faulty Predication — 46
13. Coordination and Subordination — 47
14. Sentence Clarity — 49
15. Transitions — 51
16. Sentence Variety — 53

Part Three: Parts of Sentences
17. Verbs — 56
18. Nouns and Pronouns — 63
19. Pronoun Case and Reference — 65
20. Adjectives and Adverbs — 70
21. Prepositions — 75
22. Subjects — 77
23. Phrases — 78
24. Clauses — 80
25. Essential and Nonessential Clauses and Phrases — 83
26. Sentences — 85

Part Four: Punctuation
27. Commas — 87
28. Apostrophes — 93
29. Semicolons — 95
30. Colons — 97
31. Quotation Marks — 99
32. Hyphens — 101
33. End Punctuation — 101
34. Other Punctuation — 101

Part Five: Mechanics and Spelling
- 35. Capitals — 104
- 36. Abbreviations — 104
- 37. Numbers — 104
- 38. Underlining/Italics — 106
- 39. Spelling — 108

Part Six: Style and Word Choice
- 40. Sexist Language — 122
- 41. Unnecessary Words — 125
- 42. Appropriate Words — 132

Part Seven: ESL Concerns
- 43. American Style in Writing — 137
- 44. Verbs — 138
- 45. Omitted Words — 141
- 46. Repeated Words — 142
- 47. Count and Noncount Words — 143
- 48. Adjectives and Adverbs — 145
- 49. Prepositions — 148
- 50. Idioms — 150

Part Eight: Research
- 50. Finding a Topic — 152
- 52. Searching for Information — 157
- 53. Using Web Resources — 157
- 54. Evaluating Sources — 160
- 56. Using Sources and Avoiding Plagiarism — 162

Part Nine: Documentation
- 58. Documenting in MLA Style — 165
- 59. Documenting in APA Style — 171
- 60. Documenting in Other Styles — 177

Part Ten: Document Design and Special Writing
- 61. Document Design — 178
- 62. Public Writing — 181
- 63. Writing about Literature — 183

Answer Key — 187

PART ONE
THE WRITING PROCESS

Chapter 1 Purposes and Audiences

1 a. Purpose

For each of the purposes for writing listed below, create a situation in which you might write with that purpose in mind. An example of a situation for the first purpose, "Summarizing," is given here.

EXAMPLE

Summarizing:
>I work for a consumer finance company, and my job is to look over an applicant's credit history: credit report, information from the application, etc. I must summarize the credit history in a single paragraph for the loan officer who will make the final decision.

1. Summarizing:

   _____

   _____

   _____

   _____

2. Defining:

   _____

   _____

   _____

   _____

3. Analyzing:

   _____

   _____

4. Persuading:

5. Reporting:

6. Evaluating:

7. Discussing/Examining:

8. Interpreting:

_____

_____

_____

_____

1b. Topic

In order to find a topic for a piece of writing, assume that an interviewer or reporter is asking you the following questions, and write your responses to those questions. An example of a possible answer to the first question is given here.

EXAMPLE

What is a problem you'd like to solve?

> Businesses downtown have a problem attracting customers since there is nowhere to park; a parking garage might help revitalize the city.

1. What is a problem you'd like to solve?

_____

_____

_____

_____

2. What is something that pleases, puzzles, irritates, or bothers you?

_____

_____

_____

_____

3. What is something you'd like to convince others of?

_____

_____

4. What is something that seems to contradict what you read or see around you?

_____
_____
_____
_____

5. What is something you'd like to learn more about?

_____
_____
_____
_____

6. What is something you know about that others around you may not know?

_____
_____
_____
_____

1c. Thesis

Narrow each of the following topics by asking yourself the series of questions listed with each topic. Then write the thesis that has developed from your answers. Analyze your thesis by underlining your topic once and your comment about it twice. An example is given here.

EXAMPLE

Topic: Carpentry

Who am I in this piece of writing?   I am someone with knowledge to share.

Who is my intended audience?   Readers of a do-it-yourself newsletter.

What is the purpose of this writing?   To inform.

What are some other conditions that will shape this writing? Length (space provided in the newsletter for this article), format, demographics of the audience (age, basic level of knowledge of carpentry).

Thesis:
<u>Building an attractive and efficient computer desk</u> can be an easy and affordable weekend project.

1. **Topic: Television**

    Who am I in this piece of writing? _____

    Who is my intended audience? _____

    What is the purpose of this writing _____

    What are some other conditions that will shape this writing? _____

    _____

    _____

    Thesis: _____

    _____

    _____

2. **Topic: Human cloning**

    Who am I in this piece of writing? _____

    Who is my intended audience? _____

    What is the purpose of this writing? _____

    What are some other conditions that will shape this writing? _____

    _____

    _____

    Thesis: _____

    _____

    _____

3. **Topic: Sports**

    Who am I in this piece of writing? _____

    Who is my intended audience? _____

What is the purpose of this writing? _____

What are some other conditions that will shape this writing? _____

_____

_____

Thesis: _____

_____

_____

1d. Audience

In order to be sure that the information you write is helpful to your readers, consider more fully and carefully the general audience you targeted for the topics in the previous exercise. Analyze each audience by answering the questions below.

1. Topic: **television**   Target audience: _____

   a. What do the readers already know about the topic, and what new information will they need?

   _____

   _____

   _____

   _____

   _____

   b. What is the audience's attitude toward the subject?

   _____

   _____

   _____

   _____

   c. What is the audience's background?

   _____

   _____

d. What tone or level of formality should you use?

2. **Topic:** human cloning   Target audience: _____

   a. What do the readers already know about the topic, and what new information will they need?

   b. What is the audience's attitude toward the subject?

   c. What is the audience's background?

d. What tone or level of formality should you use?

3. **Topic:** sports   Target audience: _____

   a. What do the readers already know about the topic, and what new information will they need?

   b. What is the audience's attitude toward the subject?

   c. What is the audience's background?

d. What tone or level of formality should you use?

Chapter 2 Writing Processes and Strategies

2a. Planning

Use the planning strategies listed below to find material you might want to include in writing about the corresponding topics. An example of one planning strategy, "Brainstorming," is given here.

EXAMPLE

Topic: Choosing a college

Planning Strategy: Brainstorming

    deciding between a state university and a private college
        reasons to go to a state university
- close to home
- in-state tuition much less than private college
- large selection of classes
- popular sports teams
- research facilities

        reasons to go to a private college
- smaller classes
- prestige
- less chance of being lost in the crowd
- even freshmen always taught by professors

checking out college flyers
checking out magazine articles on colleges
visiting college campuses
talking to friends and family who have attended college
knowing how much I can afford and not going above it
grades/acceptance
what SAT/ACT scores are required choosing a roommate

1. Topic: Paying for college

   Planning Strategy: Brainstorming

   _____
   _____
   _____
   _____

2. Topic: Comparing American dating to dating in another culture

   Planning Strategy: Freewriting

   _____
   _____
   _____
   _____

3. Topic: Rock music

   Planning Strategy: Listing

   _____
   _____
   _____
   _____

4. Topic: Violence on television

   Planning Strategy: Clustering and Branching

   _____
   _____
   _____
   _____

5. Topic: How to choose a college

   Planning Strategy: Outlining

   _____
   _____
   _____
   _____

6. Topic: Drinking and driving

    Planning Strategy: Who?—what?—when?—where?—how?—why?

    _____
    _____
    _____
    _____

2b–2c. Drafting and Organizing

Using one of the topics and planning strategies you completed in 2a, draft and organize a paper on that topic.

2d. Collaborating

Using one of the following methods, get responses from readers of the draft of the paper you completed in 2b-2c: meeting with a writing-center tutor, meeting with a small group of students in your class, or meeting with a writing group you form on your own. Ask your readers to respond to the questions below about your paper, and record their responses to each question.

1. What do the readers like about this draft?

   _____
   _____

2. What do the readers think is the main point of this paper?

   _____
   _____

3. Are there any sections that are unclear and need more explanation?

   _____
   _____

4. Does the paper fit the assignment?

   _____
   _____

5. Who is the appropriate reader or audience for the paper?

   _____
   _____

6. Are there any sections of the paper that seem out of order?

   _____
   _____

7. Are there any sections of the paper where the writing seems to digress from the topic?

   _____
   _____

8. Does the paper flow?

   _____

   _____

9. What else do your readers want to know about the paper's topic?

   _____

   _____

10. What is the most important revision to your paper your readers would suggest?

    _____

    _____

2e. Revising

Using the responses you gathered in 2d above and the revision checklist of higher-order concerns (HOCs), revise your paper for greater effectiveness.

_____

_____

_____

_____

_____

_____

_____

_____

_____

_____

2f. Editing and Proofreading

Using the checklist of later-order concerns (LOCs), edit your paper for details of grammar, usage, punctuation, spelling, and other mechanics. After editing for LOCs, proofread your paper one last time for missing words, misspellings, and format. Then write the final draft of your paper below.

Chapter 3 Paragraphs

3a. Unity

Read the sample paragraph below, and then write a short paragraph describing what aspects of unity are present or missing.

> A recent expedition to Tibet resulted in the discovery of a previously unknown species of a small horse. The horses are only about four feet tall, but they are very strong. Expedition members experienced a great deal of bad weather during their stay in Tibet. Even though the discovery of the tiny horses caused much excitement in the scientific world, the local citizens seem to take them for granted and use them as pack animals.

_____

_____

_____

_____

_____

_____

_____

_____

_____

_____

_____

3b. Coherence

Read the sample paragraph below, and then write a short paragraph describing what aspects of coherence are present or missing.

> Although many people think of the interior of Alaska as flat expanses of ice and snow, the geography of the area is actually quite varied. The interior of Alaska has long, cold winters and mild, short summers. The climate is actually quite dry, but most of the precipitation that does fall is in the winter and is in the form of snow. Twenty-three languages are spoken in the interior of Alaska. The area was home to forest dwelling bands of natives in historic times.

3c. Development

Read the sample paragraph below, and then write a short paragraph describing whether its development is adequate or inadequate and why.

> *Blade II* is a really good movie. It stars Wesley Snipes and is about vampires. It is really exciting and is a sequel to the first movie–*Blade*.

3d. Introductions and Conclusions

1. Ask several classmates how they go about writing an introductory paragraph for a longer piece of writing. Write a short paragraph describing the different ways people draft an introductory paragraph.

2. Ask several classmates how they go about writing a concluding paragraph for a longer piece of writing. Write a short paragraph describing the different ways people draft a conclusion.

3e. Patterns of Organization

For each of the patterns of organization listed below, write a paragraph using that pattern on the topic provided.

1. Pattern of Organization: Narration

    Topic: An experience that taught you a valuable lesson

2. Pattern of Organization: Description

    Topic: A place where you go to think

3. Pattern of Organization: Cause and Effect

   Topic: Why you chose your current major/course schedule

4. Pattern of Organization: Analogy

   Topic: The impact of the Internet on American society

5. Pattern of Organization: Example

   Topic: Contemporary heroes

6. Pattern of Organization: Illustration

   Topic: "Reality" television shows

7. Pattern of Organization: Classification

   Topic: Contemporary music

   _____

   _____

   _____

   _____

   _____

   _____

   _____

   _____

   _____

   _____

8. Pattern of Organization: Division

   Topic: College life

   _____

   _____

   _____

   _____

   _____

   _____

   _____

   _____

   _____

   _____

9. Pattern of Organization: Process Analysis

   Topic: Finding a job

10. Pattern of Organization: Compare and contrast

    Topic: Two places you have visited

11. Pattern of Organization: Definition

   Topic: Success

Chapter 4 Argument

4a-4c. Writing and Reading Arguments, Considering the Audience, and Finding a Topic

Choose a topic for a persuasive paper which is arguable, interesting, or of local concern. Once you have chosen your topic, decide on the appropriate audience for your topic, your purpose for the argument, the kinds of appeals you will use to make your case, and the common ground you share with your audience. If you have difficulty coming up with a topic of your own, you may choose from one of the following: smoking laws; the role of the United States as a world peacekeeper; the use of personality tests as a requirement for employment; a particular investment method as opposed to others (for example, investing in mutual funds as opposed to the stock market); or a local bond issue.

Topic _____

Audience _____

Purpose _____

Appeals _____

_____

Common Ground _____

_____

4d. Developing Your Arguments

1. Using the same topic for writing and the ideas you began planning in the last exercise, begin developing your argument by clarifying what your main point or claim is, what support you will offer for that claim, what warrants or unspoken assumptions are present in the argument, and what form of development (inductive or deductive) will be most effective.

Your claim _____

_____

The type of support you will use to convince the audience _____

_____

Warrants in the argument _____

_____

Method of development (inductive or deductive) _____

2. Consider how to avoid errors in reasoning in your argument by writing examples of logical fallacies about your topic which you would NOT use in a persuasive paper. By recognizing what you will not use as an argument, you will be better prepared to decide what you will use.

   a. Hasty Generalization

   _____

   _____

   b. Begging the Question (circular reasoning)

   _____

   _____

   c. Doubtful Cause (post hoc, ergo propter hoc)

   _____

   _____

   d. Using Irrelevant Proof to Support a Claim (non-sequitur)

   _____

   _____

   e. False Analogy

   _____

   _____

   f. Attack the Person (ad hominem)

   _____

   _____

   g. Either ...Or

   _____

   _____

h. Bandwagon

_____

_____

3. For more practice in logic, use the list above to identify the logical fallacies in the following statements or arguments. Write a label that names the fallacy; then explain how the example illustrates that fallacy.

   a. A vote for the President is a vote for our troops who are in harm's way; a vote for his opponent shows a lack of support for our troops.

   Fallacy _____

   Explanation _____

   _____

   b. This candidate had a reported income of over 6 million dollars last year. How can he possibly understand and help all the people in this country who don't have enough money to feed their children?

   Fallacy _____

   Explanation _____

   _____

   c. I'm going to vote for this candidate because everyone who cares about the values I care about is voting for him.

   Fallacy _____

   Explanation _____

   _____

   d. Assisted suicide, even though based on the desires of the person involved, should be illegal because anyone who wants to commit suicide is obviously incompetent and not capable of making rational decisions.

   Fallacy _____

   Explanation _____

   _____

   e. This candidate will not defend the country against attacks by our enemies. Throughout his career in Congress he has consistently voted against military spending.

Fallacy _____

Explanation _____

_____

4e. Organizing Your Arguments

Decide upon the most effective organization for the argument you were developing above and justify why you have chosen that pattern of organization.

Pattern of Organization _____

Justification _____

_____

_____

(Now that you have completed Exercises 4a-4e, you should be prepared to draft, revise, and edit a persuasive paper on your topic if your instructor requires it.)

Chapter 5 Visual Argument

Develop a visual argument for each of the situations described below. Remember to consider these questions as you plan each visual argument:

- What claim do you want to make? (i.e., what is your point or thesis?)
- Who is the audience and what beliefs can you assume they have?
- What shared images can you draw on that will be immediately recognizable to the audience?
- How can you make connections between the claim and the images?
- Which images will have an emotional impact without overpowering the logic of the argument?
- Do you need a speaker to contribute ethical appeal?
- Is the image sufficient to be effective, or do you need carefully chosen words to enhance it?
- Are you avoiding obvious logical fallacies?

In the spaces below, describe (and draw if you wish) the image you would use.

1. As a teaching intern at a middle school, you are asked to develop an image to persuade students not to begin (or to stop) smoking.

2. As an active student member of the Republican or Democratic party, develop a visual argument to persuade successful, college-educated twenty-something professionals who have never voted to do so in the current election.

3. As an activist in your suburban neighborhood, develop a visual argument to persuade your neighbors to have their pets neutered.

PART TWO
SENTENCE ACCURACY, CLARITY, AND VARIETY

Chapter 6 Comma Splices and Fused Sentences

The following paragraph contains some sentences that are punctuated incorrectly, some comma splices, and some fused sentences. Correct all the punctuation errors.

      Our country seems to be going through a change in attitudes. Unlike our forefathers, we Americans are now encouraged to get as much as possible as quickly as possible. This attitude is expressed on television advertising also plays on this appeal. Many see a relationship between these current attitudes and gambling on lotteries, it is easy to understand why more and more people are handing their money over to chance, fate and luck if what they are looking for is instant gratification. Statistics show this attitude is growing, in the late 80s, state lotteries grew an average of 17.5 percent annually—roughly as fast as the computer industry. High-tech advertising for the lotteries takes attention away from the fact that a player has virtually no chance of winning, advertisements focus instead on the fantasy of what it would be like to win. Surveys conducted by state lotteries show that few players have a clear understanding of how dismal their odds of winning really are, a player has a better chance of being struck by lightning than of winning a lottery.

Chapter 7 Subject-Verb Agreement

1. In the following paragraph, underline the verb that agrees with the subject.

    A business that is growing rapidly in popularity as well as in profits (is, are) selling meteorites to collectors, museums, universities, and research laboratories. The usual meteorite that dealers offer to customers (weigh, weighs) only a few grams or less, but sometimes, someone out collecting meteorites (find, finds) a piece that can weigh over 100 pounds. Either a whole large meteorite or some chunks (sells, sell) for thousands of dollars. But in some countries the government (is, are) beginning to pass laws that forbid the export of meteorites. None of these governments (want, wants) these national treasures to leave their countries. The price for meteorites (depend, depends) on the rarity of that kind. How much is known to exist (is, are) an important factor in determining the rarity of any specimen that is found. Of the three classes of meteorites, "irons"—made up of iron and nickel—are common, though they represent less than 10 percent of all meteorites that (strike, strikes) the Earth. But they survive the intense heat of entering the atmosphere better than other meteorites (does, do) and can be easily found with a metal detector. All other meteorites, which (is, are) made of stone, are the most common ones that are found. Among these there (is, are) some that may have come from nearby planets when meteorites hit their surfaces. Having a piece of Mars (is, are) a good way to start a collection.

2. As you complete the following sentences, make sure the verb agrees with the subject. For this exercise, use only present or past tense verbs. For example, you could write I am happy, or I was happy, but do not write I will be happy.

a. I _____

_____

b. We _____

_____

c. Nearly all of my sister's clothes which she wears to school _____

_____

d. The store clerks and their manager _____

e. Either Carlos or his brothers _____

f. Not only the players but also the coach _____

g. What I want to explain _____

h. To feel successful _____

i. Each _____

j. Some of the examples _____

k. Some of the assignment _____

l. The committee _____

m. Cryogenics _____

n. The scissors _____

o. Acme Enterprises _____

p. Her idea (complete this sentence using a linking verb) _____

q. Magazine articles (complete this sentence using a linking verb) _____

r. There _____ an interesting book on the history of television in the library.

s. There _____ many books on the history of television in the library.

t. It _____ his beliefs that formed the basis for his argument.

u. They are the people who _____

v. She is the person who _____

w. Mr. Baker is one of those instructors who _____

Chapter 8 Sentence Fragments

1. The following paragraph has both fragments and complete sentences. Identify each sentence as either a fragment or a complete sentence.

   (1) A fierce battle has developed in American schools over the question of using English in the classroom. (2) Historically, immigrants coming to America were expected to learn English and to be taught in English. (3) Though that did not mean that they had to reject their original culture or language. (4) But for many, using English in school meant that their ethnic legacy would no longer dominate. (5) That it would disappear as they became fluent English speakers. (6) As a result, the emphasis on English as the dominant language was seen by many as discriminatory. (7) While others see this emphasis as the only way to have a common American culture that all children can share. (8) One answer to this has been bilingual education in which children who speak another language are taught academic subjects in their native language. (9) This transitional program which helps them to move along through school while they acquire English. (10) Bilingual education being the answer for some who want to preserve children's culture and provide schooling for them as they learn English. (11) Although others see it as a failed system because it segregates limited-English children into separate classrooms and dooms them to unskilled jobs where competence in English is less important. (12) They point out that children need to learn English in order to get better jobs, go on to college, and become full members of American society

   1. _____
   2. _____
   3. _____
   4. _____
   5. _____
   6. _____
   7. _____
   8. _____
   9. _____
   10. _____

11. _____

12. _____

2. Revise the sentences you identified above as fragments so that they are complete sentences. Write the number from the list above; then rewrite the sentence.

_____
_____
_____
_____
_____
_____
_____
_____
_____
_____
_____
_____
_____
_____
_____
_____
_____
_____
_____
_____

Chapter 9 Dangling and Misplaced Modifiers

9a. Dangling Modifiers

1. Underline the dangling modifiers in the following paragraph.

   The great memorials in the nation's capital are slowly deteriorating. Having faced the elements for seven decades, even casual tourists can see that the Lincoln Memorial is wearing down. Water is causing the steel structures embedded in the concrete slabs to deteriorate. At the Jefferson Memorial, which is built on clay fill, the steps are starting to pull away from the base of the monument, and there are cracks appearing in the dome. Being washed daily, details in the marble carvings of both monuments are falling off. Air pollution and acid rain are also contributing to the gradual disintegration. To find a solution, constant studies by the National Park Service are being conducted, but the caretakers of the monuments are pessimistic. Exposure to the elements, they say, causes some deterioration that cannot be stopped. Looking more like ruins than buildings, it is hard to watch these noble monuments wear away.

2. Rewrite the sentences with dangling modifiers that you identified in the exercise above, eliminating the modifier errors.

   _____
   _____
   _____
   _____
   _____
   _____
   _____
   _____
   _____
   _____

9b. Misplaced Modifiers

1. In the following paragraph, underline the misplaced modifiers.

   The Food and Drug Administration (FDA) claims that it has found tiny amounts of drug residue in milk samples in a report issued recently. The level of contamination is so low that there is no real safety problem in the nation's milk supply, but a ban has been put on one drug, sulfamethazine, found in a few samples, to prohibit its use with beef cattle and pigs by farmers. The FDA has already banned the use of sulfamethazine with milk cows. However, there is still concern that some dairy farmers or veterinarians use the drug improperly among members of Congress, even though recent testing of milk samples only found a few traces of sulfa drugs in a large number. The FDA is now calling on dairy farmers and veterinarians to eliminate all illegal uses of veterinary drug products. Farmers are already almost all complying with this order.

2. Rewrite the sentences you identified above, placing the modifiers appropriately.

9c. Additional Modifier Exercises

Identify each sentence below as having a misplaced or dangling modifier and rewrite it, correcting the problem

1. Using primitive tools, scientists have discovered Cro-Magnon man was human.

2. Bob stunned the pit bull with a large flashlight.

3. Walking into the house's entryway, the lamp glowed warmly.

4. Please choose a magazine on the enclosed card.

5. Sue found an emerald woman's bracelet.

6. Approaching the city from the west, the sports arena is quite a sight.

Chapter 10 Parallel Construction

1. In the following paragraph underline all parallel constructions and rewrite any nonparallel forms so that they are constructed in parallel form.

   Recently, two pilots, one in a 175-seat commercial airliner and with the other one in a small, twin-engine corporate jet, were barreling toward each other. On the instrument panel of the commercial jet, a small air traffic screen flashed a yellow circle and a voice announced, "Traffic." As the yellow circle approached the center of the screen, it changed to a red square. The voice said loudly, "Climb, climb." Noticing the red square and as he pulled up, the pilot saw the other craft fly past several hundred feet below. The voice that called out the warning was not the copilot but the latest audiovisual aid to arrive in cockpits of planes, a traffic alert and collision avoidance system. The Federal Aviation Administration has issued an order that this system must be installed in 20 percent of all large commercial planes by next year and to install it in the rest of the large commercial planes using American airspace within the next three years. The system works by computing the distance between planes, warning planes when they get within six miles of each other, and then to decide which plane should climb and which should descend to avoid a collision.

2. Write a paragraph of at least five sentences in which there are parallel constructions in at least four of the sentences. You may want to write about air safety or airport safety.

Chapter 11 Consistency/Avoiding Shifts

1. The following paragraph has a number of inconsistent shifts in person and number, in verb tenses, in tone, in voice, and in discourse. Rewrite the paragraph on the lines below so that it avoids any inconsistent shifting.

   High school proms used to be dances where students dressed up in suits and dresses and celebrated their coming graduation, but now you have to spend big bucks for a tuxedo or elegant formal dress. Proms have become big business as formalwear shops and limousine services offer their services in advertising campaigns. Tuxedo rental shops across the country report that proms, not weddings, account for the major proportion of its business. One local shop owner said that he used to look forward to summer as his busy season because of weddings, but now I make more money in the spring because of high school proms. The typical expenses now include the tuxedo rental, prom tickets, corsage, dinner, and limousine rental. You can easily spend $300 on the dance, and that was just for the basics. Promgoers can spend additional funds for the latest fashions in tuxes, and photos can be bought for $50 or more. Even when the prom is over, there were other expenses. Some kids go away for the whole weekend, often to a resort hotel. So you have to throw in the costs for a hotel, and if parents are along as chaperones, it is necessary to add in the cost of their rooms and meals too. High school graduates insist that this is all necessary as a rite of passage, but the expense is not appreciated by parents who often have to cough up at least part of the funds.

2. Write a paragraph of at least five sentences in which you pay special attention to consistency in pronoun person and number, verb tense, tone, and voice. You may want to write about high school proms.

Chapter 12 Faulty Predication

1. The following paragraph has some sentences with faulty predication. On the lines below, rewrite these sentences so that they are correct.

    With the present concern for the environment, some companies are trying to increase their sales by advertising their products as environmentally safe. The makers of some plastic trash bags, for example, are claiming that their plastic is degradable. The reason for the claim is because there are additives that cause the product to break down after prolonged exposure to sunlight. Biodegradability, they assert, is when there is photodegradability, a breakdown by sunlight. But since most trash bags are buried in landfills, the benefits of photodegradability are questionable. Thus, the government has stated that one way to improve deceptive advertising claims is when they eliminate false or misleading information.

    _____
    _____
    _____
    _____
    _____
    _____
    _____

2. Finish the sentences for each of the subjects listed here. Pay special attention to avoiding faulty predication.

    a   The reason is _____

    b   Horror is _____

    c   His excuse was _____

    d   Her proposal was _____

Chapter 13 Coordination and Subordination

The sentences in the following paragraph can be improved by adding more coordination and subordination. Rewrite the paragraph so that there is more coordination and subordination, but avoid excessive or inappropriate forms of adding clauses together.

      Americans have become videotape addicts. Over 75 percent of all American households now have videocassette recorders, so hotels hope to make their guests feel at home by providing equipment for watching videotapes. Some of the largest hotel chains in the country have added videocassette recorders in the rooms which can be used by renting videocassettes from shops in the hotel where there is a good selection of recent movies for guests to view in their rooms. In these hotels guests can show their room key and can rent a movie at a reasonable rate. Other hotels don't want to bother with stores in the lobby, and they are exploring a different option, so they are adding automated video dispensing machines in their lobbies that hold hundreds of titles and have new releases as well as standard favorites. One hotel chain has a lot of large business conventions, and it has investigated another approach. It is offering to distribute to guests videotapes that the corporation holding the convention wants its participants to see since corporations like this because the videotapes can convey some of the key ideas being presented at the convention. Videotape players in hotel rooms may soon be standard equipment, just as television sets were added years ago when Americans became addicted to televisions so that they expected to watch televisions in their hotel rooms.

___
___
___
___
___
___
___
___
___

Chapter 14 Sentence Clarity

1. The following paragraph has some sentences with clarity problems. Revise the paragraph to improve sentence clarity.

   To jot handwritten notes onto an electronic pad may be a new approach to using computers. A special pen that projects a narrow light beam onto the pad is the way this is done. For computer users who have had to rely on entering data into a computer by means of a keyboard, it is a major step forward in using computers. Not having to type is less distracting to many people who are not skilled typists but whose computer usage is high. It has been the goal of computer developers to rid the computer of keyboards for the last twenty-five years. But there has been insufficient development in the field of character recognition as this is necessary for the elimination of keyboards. Teaching computers to read through optical character recognition is one way to eliminate the keyboard. Recognition of the human voice is another way for computers to cease to rely on keyboard inputting of data. Speech recognition is not advancing as rapidly as some computer developers would like, and it is not likely that the near future will see computers we can converse with. We hardly have not appropriate technology that is this advanced. More promising is the ability of the computer to scan images. Already character recognition machines are being used in offices where pages of printed material are scanned by machines. Also under consideration by computer developers are electronic gloves that could be used by people to point to areas of the screen. There is hardly no limit to what will be coming next in computer development. The doing away of the keyboard will result in saving time and in the elimination of all those typos.

2. Write a paragraph of at least five sentences in which you draw arrows from old information in sentences to new information. Use positive instead of negative, avoid double negatives, and pay special attention to using verbs instead of nouns. Underline any passive voice verbs and be sure that they are appropriate. You may want to write about your experience in using computers.

Chapter 15 Transitions

1. Rewrite the following paragraph by using transitions to build bridges between the sentences and parts of sentences. Use underlines to indicate the transitions added in your revised version.

   My father really needs help. He is a workaholic. He works nearly twelve hours a day. Doctors have told him it is not only bad for his health, but it is affecting his family. He has worried my family to no end. My family has tried to help him. He has not acknowledged the problem. He has a tolerance that goes through the roof. He sneaks out of the house before anyone is awake. He gets home after most of us have finished dinner and are getting ready for bed. He works at home. His computer is connected to the one at his office. His work calls him at home. He has a pager and a cell phone. Things may be getting worse. Dad accepted a promotion to regional director.

   _____
   _____
   _____
   _____
   _____
   _____
   _____
   _____
   _____
   _____
   _____

2. Write a paragraph of at least five sentences in which you use transitions in at least four of the sentences. Underline your transitions. You may want to write about your experience with delays in grocery store checkout lines.

   _____
   _____

Chapter 16 Sentence Variety

1. The following paragraph has very little sentence variety. Revise the paragraph by using a combination of strategies to add variety.

   Animal rights activists are known primarily for their campaigns against fur coats and the use of animals in laboratories. They are also campaigning against rodeos. Protesters say that rodeo animals are being mistreated. They say the animals are starving and live a life filled with pain and suffering. Animal rights activists also say that rodeo horses buck because they are in pain. Rodeo and circus owners say that this is not so. These animals are well fed and comfortable. These animals would be going to slaughter if they were not used as show animals. Handlers explain that they have a healthy respect for the size and power of these animals. The animals are treated like star athletes. They enjoy performing. Animal rights activists point to the use of cattle prods and bucking straps to get the animals moving. Rodeo owners argue that even ranchers have to use electric prods to get herds moving. Ranchers used pitchforks before prods were used. Calf roping is also condemned by animal rights activists. Roping breaks calves' necks and can also snap vertebrae and legs. Calf roping has been banned in some states. Eliminating calf roping can result in the overall elimination of rodeo shows. Rodeo owners protest that breaking of calves' necks, bones, and vertebrae just doesn't happen. Rodeo owners invite activists to come to rodeos and see for themselves.

2. Find a paragraph you have written previously that you think needs more sentence variety. Rewrite the paragraph here so that you have combined sentences; added words, phrases, or clauses at the beginning of the sentence; or changed some sentences to dependent clauses within the sentence.

3. The sentences in the following paragraph are all short, simple sentences, so the paragraph is choppy and unsophisticated. Improve the style and sense of the paragraph by adding sentence variety, combining the sentences with coordination and subordination, and adding transitions as needed.

   Gardening can be a worthwhile activity. The gardener can produce food for her family. She can also help to save the environment and prevent the overuse of landfills. One way to do this is by mulching. Mulching is actually recycling. The gardener can recycle grass clippings and leaves. The clippings are from mowing the yard. The leaves have to be raked in the fall. The clippings and leaves are usually gathered in plastic bags. The bags are taken to the landfill by the waste collection companies. The gardener can use scraps from the kitchen. These scraps are usually discarded into plastic garbage bags also. The gardener can collect clippings and leaves in a corner of the garden. She can add scraps of food to the clippings and leaves. She can mix these all together to provide organic mulch. This mulch can be spread on the garden. The mulch will provide nutrients for the soil. The nutrients will cause the garden to produce bigger and better vegetables. The gardener reaps many rewards from her activities.

PART THREE
PARTS OF SENTENCES

Chapter 17 Verbs

17a. Verb Phrases

1. Underline the verb phrases in the following paragraph.

    Some scientists are saying that a buildup of carbon dioxide and other greenhouse gases in the atmosphere will cause global warming. But another group of scientists argues that we should study the data more carefully before any firm conclusions are drawn. While scientists generally agree that an unchecked accumulation of greenhouse gases will cause changes, no one knows when it will start, how much will happen, or how rapidly it will occur. The most widely accepted estimate is that there will be a rise in the earth's average temperature as early as 2050. This could bring rising sea levels and severe droughts in some areas. But no one knows yet how clouds and the ocean's ability to absorb heat will affect this. When scientists understand this better, projections can be revised.

2. Write a paragraph of at least five sentences, and underline the verb phrases. You may want to write more about the greenhouse effect and global warming.

   _____

   _____

   _____

   _____

   _____

   _____

   _____

17b. Verb Forms

1. In the following paragraph, underline verb forms that are part of a verb phrase with one line and verb forms that are used alone with two lines.

   The evidence that global warming has started is not very strong. Some scientists believe that the concentration of carbon dioxide has increased over 25 percent since the early 1800s, but other scientists point to the fact that the average global temperature has risen by no more than a half degree Centigrade. Even that rise is questionable since there was a cooling period from 1940 to 1970 that caused forecasters to predict a return to the ice ages. Therefore, to act on predictions by passing laws that restrict or ban the use of fossil fuels may be hasty, but conserving energy, banning harmful chlorofluorocarbons, and planting more trees to absorb carbon dioxide from the air seems sensible. Many industries are also acting more responsibly and are reducing hazardous emissions from their factories.

2. Write a sentence using each of the following verb forms not as a part of the verb phrase but elsewhere in the sentence.

   a. -ing verb

   b. -ed verb

   c. to + verb

17c. Verb Tense

1. Underline the correct verb tenses from the choices given in the following paragraph.

   In 1987, the FBI (opened, has opened, was opening) a $1.5 million mock-up of a small town in Virginia named Hogan's Alley. Presently, the town's population (has been, will be, is) about 200. Generally, Hogan's Alley (has looked, looks, is looking) like any peaceful little American town, but often the quiet (will be shattered, is shattered, had been shattered) by the sound of a shotgun or squealing tires. Then prospective G-men who (had been sitting, sat, will sit) outside the post office jump in their cars and race after the "criminals." None of this (will be, is, is being) real because Hogan's Alley (is, has been, had been) a training academy for those who (will want, want, will have wanted) to become FBI agents. Next year over 500 trainees (will have attended, attend, will attend) classes in frisking, lectures on handcuffing, and seminars on interrogating witnesses. The FBI (started, was starting, had started) Hogan's Alley because it wanted its agents to have more true-to-life experience before they go out and deal with dangerous criminals. So now every day there (will be, have been, are) mock bank robberies, kidnappings, and drug busts. The "criminals," however, are actors, part-time students, retirees, off-duty policemen and firemen, and anyone else who (has passed, will pass, had passed) the rigorous screening tests.

2. The following paragraph is written in present tense. At the beginning of the paragraph, add the words "Last year" and change the verbs in the rest of the paragraph so that it is in past tense. Be especially careful when changing irregular verbs.

   In Hogan's Alley, the FBI's mock-up village for prospective FBI agents, the trainees take part in an intensive 14-week training course. About 46 agents move through the program which begins with lessons in surveillance. Trainees track a suspect from his home as he drives to a shopping mall and sells a small bag of phony cocaine. Next comes a class in simple arrests, when agents-in-training burst into a fleabag motel and take an unarmed burglar as he lies in bed. Agents learn how to frisk suspects, and then they read them their rights. Instructors, who bring their own experience as former FBI agents to the teaching, carefully give advice to each trainee and provide pointers on how to handle details such as slipping on handcuffs during a struggle. Nine weeks into the course, trainees get some practice in arresting an armed felon. Their final lessons take place in a courtroom, where they face a team of actors who pose as a tough team of defense lawyers. Those trainees who do not flunk out of the course become FBI agents, and the actors, who play drug peddlers, burglars, and other felons, say that seeing what it is like being on the wrong end of the law reminds them that they do not want to become criminals themselves.

17d. Verb Voice

1. The following paragraph contains both active and passive voice verb phrases. Indicate the voice by underlining "active" or "passive" in the parentheses.

   Like many cities built on a river, Kansas City is divided (active, passive) between two states. The biggest single municipality is Kansas City, Missouri, but over one-half of the population of the metropolitan area lives (active, passive) in various suburban cities on the Kansas side. Problems are created (active, passive) when it comes time to finance projects that will benefit (active, passive) both sides of the state line. In the 1990s, cities on both sides of the state line agreed (active, passive) to a bi-state sales tax to be used to renovate Kansas City, Missouri's Union Station. Advocates said (active, passive) the new station would bring (active, passive) in tourism that would benefit both sides of the state line. Opponents claimed (active, passive) Kansas City, Missouri, was just trying to dip (active, passive) into the pockets of the more affluent suburbs on the Kansas side. In the end, a compromise was reached (active, passive). The bi-state sales tax was passed (active, passive), but all contributing cities had (active, passive) a say in its use and the tax had (active, passive) a 5-year duration. It was not renewed (active, passive).

2. Examine the sentences in the previous paragraph that you identified as having passive verbs. Rewrite them so that the verbs are in active voice or explain why the passive voice is more appropriate and less awkward.

   _____

   _____

   _____

   _____

   _____

3. Write a brief paragraph with at least five sentences in which you appropriately use three or more active verbs and appropriately use two or more passive verbs. Underline the active verbs once and the passive verbs twice. You may want to write about American inventions that interest you.

17e. Verb Mood

1. Some of the verbs in the following paragraph are declarative (express a fact), some are subjunctive (express some doubt or something contrary to fact), and some are imperative (express a command). Underline the mood used for each.

   Singles bars and dating services are thriving (declarative, subjunctive, imperative), but there are always new approaches. In commercial dating services, one new approach that may be cheaper (declarative, subjunctive, imperative) than the standard videotaped interviews is the lunch-date service. For less than $50 a month, the company promises (declarative, subjunctive, imperative) three lunch dates a month. People are paired on the basis of simple criteria gathered from brief interviews that might last (declarative, subjunctive, imperative) less than five minutes. The company sets up the lunch date, and the participants take it from there.

"Meet (declarative, subjunctive, imperative) new people," says the advertising brochure, "and enjoy (declarative, subjunctive, imperative) some interesting little restaurants." Since the Census Bureau puts (declarative, subjunctive, imperative) the number of single adults at 66 million and growing, these new twists on dating services may prosper (declarative, subjunctive, imperative).

2. Write sentences using each of the verb moods listed below.

   a. declarative

   _____

   _____

   b. subjunctive

   _____

   _____

   c. imperative

   _____

   _____

17f. Modal Verbs

1. The following paragraph includes modal verbs that express ability, a request, or an attitude (such as interest, expectation, possibility, or obligation). Underline the correct meaning from the choices given in parentheses after the modal verbs.

   When American products began appearing in Moscow, city officials worried that the signs and advertising for these products might make (intend to make, could possibly make, need to make) Moscow look less Russian. A law recently passed in Moscow warns that all stores and businesses must display (are capable of displaying, expect to display, need to display) signs in Russian or at least change them into the Cyrillic alphabet. This has caused many businesses to contact the city inspector because of questions they have. For example, one businessman wondered whether he should change (has the ability to change, asks to change, is obliged to change) the letters in the label for the Puma running shoes that he sells. He was concerned that changing the letters from English to Russian might make (has the possibility of making, intends to make, requests to make) the shoes less well-known. The problem became confusing because the new law does not forbid foreign words but does require the Russian sign to be bigger than the one in English. Different stores found different solutions. An American cosmetics company, Estee Lauder, announced that they will put (are capable of putting, strongly intend to put, are asking permission to put) one sign with the Russian alphabet on their awning and another sign with the English alphabet in the windows. Despite all the questions and worries, the inspector in charge of enforcing this rule will be very strict about checking on foreign signs.

2. Write a brief paragraph with at least five sentences in which you use four or five modal verbs (shall, should, will, would, can, could, may, might, must, ought to). Underline each modal verb that you use. You may want to write about problems in translating words from one language to another or about the difficulties of having to use the English alphabet to write words from a language that uses a different alphabet.

Chapter 18 Nouns and Pronouns

18a. Nouns

1. In the following paragraph, make the necessary corrections to the nouns that need the appropriate -s and -es plural noun endings and to the nouns that are missing possessive markers. Underline your corrections.

   Sports fanatics are spending large sums of money these day to purchase sports memorabilia. In fact, says one of the industrys spokesperson, it is a $100 million-a-year obsession. Buyers look for scorecards, stadium seats, autographed baseball bats, and even children clothing that belonged to old-time greats such as Babe Ruth. Ruth personal letter to a fan recently sold for over $7000. Since Joe Lewis, the famous boxer, kept many of his old mouthpiece, collector are paying over $2000 for boxed sets of Lewis's mouthpieces. There is even a magazine entitled *Sports Collectors Digest* which keeps collector informed about many of the auction and the availability of various item of interest. A faculty member in the history department of a major American university has begun a study of this phenomenon and reports that yearly increases in prices are staggering. For example, he says, a nineteenth-century baseball card of an obscure Hall of Fame pitcher, Tim O'Keefe, which was worth about $750 two year ago, recently sold for over $30,000. O'Keefe's other memorabilia are expected to soar in value as well. Sports collectible are now big business.

2. Write a paragraph of at least five sentences and include nouns with -s and -es plural endings and nouns with correct possessive markers. Underline each such noun that you use. You may want to write about collecting sports memorabilia.

18b. Pronouns

In the following paragraph, underline all the pronouns, including the personal pronouns, demonstrative pronouns, relative pronouns, interrogative pronouns, indefinite pronouns, possessive pronouns, reflexive pronouns, and reciprocal pronouns.

The business of sports collectibles has become so profitable that <u>it</u> has attracted con artists <u>who</u> manage to forge and sell bogus items. The forgeries have become such big business, in fact, that many con artists help <u>one another</u> and have developed a large network of bogus items. <u>These</u> items include fake Joe DiMaggio signatures and imitation press box pins. <u>Anyone</u> <u>who</u> is in the memorabilia business can spot <u>these</u> forgeries and fakes, but sports fans are often too enthusiastic about a purchase to take the time to have <u>their</u> purchases checked by experts. As a result, <u>they</u> often allow <u>themselves</u> to be conned. A lot of junk from people's attics is also cluttering the market. When <u>someone</u> finds yellowed pages from 1929 sports sections in <u>her</u> scrapbook, <u>she</u> may think <u>she</u> has an expensive treasure. Flea markets overflow with <u>these</u> antiques <u>which</u> may or may not be worth <u>anything</u>. But if a buyer is willing to pay large sums of money, <u>who</u> can say whether <u>that</u> autographed photo or threadbare jersey is or is not a treasure worth collecting? With a market where buyers pay $20,000 for <u>one</u> of Lou Gehrig's old bats, <u>it</u> is worth cleaning out <u>those</u> old attics and scrapbooks.

Chapter 19 Pronoun Case and Reference

19a. Pronoun Case

1. In the following paragraph, underline the correct pronoun in the parentheses.

When historian Page Smith began to ask himself just what kind of education was being provided by American universities, he started with a list of questions. (His, His') answers can be found in (his, his') book entitled *Killing the Spirit*. (Them, Those) readers looking for encouraging conclusions will not find (they, them) in this report as Smith criticizes professors' lack of interest in teaching and (their, theirs, their') tendency to spend too much time writing books and articles that do not contribute anything useful to mankind's betterment. Smith finds (them, those) academics who look down on teaching as less important than research to be at fault. For (he, him), the student revolt of the 1960s and early 1970s was a time when students were looking for answers to important questions but got no help from their professors. There was no dialogue between their professors and (they, them) that addressed questions about meaning and values in life, says Smith. However, community colleges and small, independent colleges still concern (theirselves, themselves) with the needs of students. University administrators (who, whom) are more interested in budgets and specific agendas will not find answers in Smith's book, and readers will wonder (who, whom) his real audience is. To (who, whom) should (we, us) look for answers?

2. Write your own sentences below correctly using the pronouns listed.

   a. them

   _____

   _____

   b. whom

   _____

   _____

c. his

d. yours

e. theirs

f. ours

g. her

h. me

i. who

j. your

k. us

_____

_____

l. I

_____

_____

3. In the sentences below, underline the correct pronoun.

  Fall break was over, and the research assignment was due the next day, so Ross asked Lucy if she wanted to go to the library with Amy and (he, him). The instructor had told the class that the students (who, whom) made high grades on the midterm exam would be those (who, whom) had read the assignments and completed all the homework. When the girls and (he, him) reached the reference room, they saw several classmates (who, whom) they could work with. They asked their friends, "(Who, Whom) has finished the assignment?" They were surprised that all the other students were finished and headed to Starbucks. It was clear that a long night lay ahead for the girls and (he, him). The next day, Ross asked the teacher, "Please give Lucy, Amy, and (I, me) one more day to finish the research assignment." But the teacher said it wouldn't be fair to the rest of the students (who, whom) had all worked hard, so Ross and (they, them) learned, yet again, not to procrastinate.

19b. Pronoun Reference

1. In the following paragraph there are some pronoun reference problems. Underline all the pronouns that do not clearly or correctly refer back to specific nouns.

  When Benjamin Franklin discovered electricity in thunderclouds, he sparked a controversy that still has no clear answer. How do clouds become electrified in the first place? To this day they haven't been able to adequately explain how it contains such incredible amounts of electricity that a stroke of lightning can contain about 100 million volts. One researcher who wants to find some answers flies his plane into storms to measure electric fields and ice particle changes. It's bumpy work, he says, especially if there are large hailstones because it can damage the plane and the measuring equipment. On one trip they noticed that in sections of clouds where water and ice mix, the measuring devices picked up indications of strong charge separation. The answer may be that in a certain temperature range, the temperature can cause the charge separation. Another factor may be a kind of soft hail called "graupel," pea-sized particles that can look like miniature raspberries. They form when droplets of supercooled water collide, freezing together instantly. Ice crystals then bounce off the growing graupel, building up a charge from the friction just as you build up a static electricity charge when you scuff your feet across the carpet. When they are carried to different parts of the cloud, the result is a separation of the positively and negatively charged particles. Then, when the electrical difference between the ground and the sky becomes great enough, everyone should haul his or her kite in.

2. On the spaces below, rewrite the sentences from the paragraph above that have pronoun reference problems so that all the pronoun problems are eliminated.

3. Write a paragraph of at least five sentences with correctly used pronouns in every sentence. Underline each pronoun and draw an arrow back to the noun to which it refers. You may want to write about electricity or thunderstorms.

Chapter 20 Adjectives and Adverbs

20a. Adjectives and Adverbs

1. In the following paragraph, underline all the adjectives and adverbs and correct the errors in adverb and adjective forms by writing the correct form above the line.

   For many people, the crossword puzzle in the daily paper is one of life's little pleasures. Some say it is also sureone of life's frustrations. While puzzle books conveniently include the answers in the back, most newspapers print the answers the next day. Now there is a more quicker answer. Some companies have an automated solution. Readers can dial an 800 service for instant answers. This service is free to callers but is paid for by advertisers who sponsor each day's puzzle. An advertiser can run a small advertisement by the puzzle and can include a ten-second message that callers will hear before the answers are given. *The New York Times* handles this more differently. For their puzzles, callers pay for all requests for clues. These services are a major breakthrough for frustrated puzzle-doers who are used to waiting until the next day.

2. Write sentences correctly using each of the adjectives or adverbs below.

   a. nice

   b. nicely

   c. sure

   d. surely

e. used

f. good

g. well

h. rapid

i. rapidly

j. real

k. really

l. so

20b. A, An, The

In the following paragraph, articles (a, an, the) may or may not be needed before certain nouns. Underline the correct answer in parentheses. Notice that one possibility is that no article is needed; if that is the case, underline that option.

After winding down twenty miles of dirt road in Mtunthama, Malawi, (a, an, the, no article) visitors will come upon the well-kept lawns and gardens of (a, an, the, no article) Kamuzu Academy, one of (a, an, the, no article) Africa's most unusual schools. Here, on four hundred acres of well-trimmed lands is (a, an, the, no article) school dedicated to classical scholarship. (A, An, The, no article) school was founded by President Hastings Kamuzu Banda, the ruler of Malawi since it won its independence from (a, an, the, no article) British in 1964. Originally, Dr. Banda studied in (a, an, the, no article) South Africa, (a, an, the, no article) United States of America, and (a, an, the, no article) Britain, where he acquired (a, an, the, no article) medical degree and (a, an, the, no article) love of Latin and Greek, as well as a strong attraction to the classical emphasis of the elite British schools. When Dr. Banda returned to Malawi, he wanted to copy (a, an, the, no article) architecture and curriculum of Eton and other British boarding schools. Some critics in opposition to Dr. Banda say that (a, an, the, no article) academy is not appropriate in their country where ordinary schools often do not have (a, an, the, no article) textbooks and where more than two-thirds of (a, an, the, no article) population are illiterate. However, defenders of the school point out that the school is not elite in choosing its students. Children are accepted without regard to their family's wealth or position. Each year 35,000 students take (a, an, the, no article) exam to try to gain entrance to the school which accepts about eighty new students (a, an, the, no article) year. Once accepted, all students are required to take four years of (a, an, the, no article) Latin and four years of (a, an, the, no article) ancient Greek, along with (a, an, the, no article) English, (a, an, the, no article) mathematics, and (a, an, the, no article) history course about Africa. Most graduates go on to university and then take jobs in (a, an, the, no article) Malawi civil service.

20c. Comparisons

Write sentences in which you use the correct comparative form of the word listed for the items given.

1. comparative form: large

    items compared: two colleges

2. comparative form: large

    items compared: eight colleges

3. comparative form: exciting

    items compared: two movies

4. comparative form: enjoyable

    items compared: three books

5. comparative form: user-friendly

    items compared: all computer software programs tested

6. comparative form: good

    items compared: two music CDs

7. comparative form: good

   items compared: three music CDs

   _____

   _____

8. comparative form: bad

   items compared: all meals you have ever eaten

   _____

   _____

Chapter 21 Prepositions

1. In the following paragraph, underline the correct choice for the preposition that should be used.

    Americans used to pour ketchup over so many foods that it was America's favorite condiment. Now salsa has replaced ketchup as the favorite (among, between) the two (at, in, on) American food markets. Salsa, which means "sauce" (at, in, on) Spanish, is defined as any fresh-tasting, chunky mixture, usually made with tomatoes, chilies, onions, and other seasonings. Although salsa used to be associated (to, with, for) Mexican and southwestern cooking, it is being used for a variety of foods not particularly Mexican. The popularity (of, about, to) salsa is apparently part of the current food trend as Americans become more interested (about, on, in) spicier foods. In 1988, only 16 percent of households (at, among, in) America bought salsa. (In, At, For) two years, that figure was up to 36 percent, and the market continues to grow (in, at, by) a very fast rate. A marketing information company notes that salsas and picantes—which are different (from, than) salsas because they are thinner—account (on, to, for) about two-thirds of this market, a category that also includes taco and enchilada sauces. As the market expands, so do the choices. The simplest salsas are based (about, around, on) chopped tomatoes and chilies, and the types of chilies determine how hot the particular type of salsa is. Cookbooks with a variety of recipes indicate the preference of some people to make their salsas (in, at, to) home. Salsas are one of the few popular snack foods that are fat-free or nearly so, and many are made without preservatives, two characteristics that may contribute (on, to, around) their popularity now that people are interested (about, in) eating healthier. Riding on this wave of popularity are the new fruit salsas, made with peaches, pineapples, and so on, and vegetable salsas, made with pinto beans, corn, and black-eyed peas. Other varieties will surely appear as the market expands (in, on, at) the future.

2. Write sentences in which you correctly use the prepositions listed.

    a. compared to

    _____

    _____

b. compared with

c. independent of

d. similar to

e. different from

Chapter 22 Subjects

1. In the following paragraph, underline the subjects of all the verbs.

    Because television <u>viewers</u> of sporting events like to follow the actions of specific players, a <u>cable company</u> in New York is trying out a new subscription service. If <u>it</u> is successful, <u>it</u> is likely to become available to everyone: <u>Cable subscribers</u> will be able to tune into a tournament and pick the <u>player</u> <u>they</u> want to follow. Using a wand similar to a TV remote control, <u>viewers</u> sitting at home will be able to switch from conventional network coverage to a <u>camera</u> that focuses solely on the preferred player. Instant <u>replays and a list</u> of statistics on selected pros will be other selections that the <u>service</u> will provide. This interactive <u>technology</u> will allow viewers to customize programming. While the present <u>programming</u> is only a test of the technology, some cable <u>companies</u> foresee customers being able to sign up for interactive television just like other cable services. When will this <u>service</u> be available for everyone? <u>It</u> is too soon to know for sure, but <u>there</u> is certainly a large demand for such services.

2. Write a paragraph of five or more sentences and underline the word or words that are the subjects of all the clauses in your sentences. You may want to write about television viewing, cable services, or active vs. passive television viewing.

    _____
    _____
    _____
    _____
    _____
    _____
    _____
    _____
    _____

Chapter 23 Phrases

1. For each of the following sentences, choose the letter which best indicates the function of the underlined phrase. Write the letter in above the phrase.

    a. subject of the sentence

    b. tells more about the subject

    c. is the verb phrase

    d. gives added information about the verb

    e. completes the subject

    f. gives added information about another element in the sentence

    1. Marketers <u>have been working on</u> ways to improve aerosol dispensers so that they do not spray the harmful propellant into the atmosphere.

    2. One new product, <u>an alcohol-free hair spray</u>, cuts emissions of chemical pollutants by roughly 60 percent.

    3. For its propellant, this new hair spray uses dimethyloxide, a <u>nondamaging chemical</u>.

    4. Reformulating a proven compound <u>can be expensive</u> for manufacturers.

    5. Another solution to the problem is <u>a propellant-free container</u> with a rubber expanding sleeve.

    6. <u>The rubber's propensity to contract</u> provides the pressure to produce a spray when the nozzle is pressed.

    7. This new spray dispenser is already being widely used <u>with great success</u>.

2. Write your own sentences with phrases that serve the functions listed here, and underline the phrases you have used fulfilling those functions.

    a. phrase that is the subject of the sentence

    _____

    _____

    b. phrase that tells more about the subject

    _____

    _____

c. phrase that is the verb phrase

d. phrase that gives added information about the verb

e. phrase that completes the subject

f. phrase that gives added information about another element in the sentence

Chapter 24 Clauses

24a. Independent Clauses

1. In the following paragraph, underline the independent clauses.

    If you are a magician and like highly structured organizations, you may want to join either the International Brotherhood of Magicians or the Society of American Magicians. If, however, you are a magician who lives in New York City and likes to gather informally with other magicians, you should visit Reuben's Restaurant on 38th Street and Madison Avenue any Saturday afternoon. There will be professional and amateur magicians sitting in the back room, and they will be swapping tricks or polishing their routines on each other. People come in with jumbo interlocking rings, decks of cards, and brightly colored scarves in their pockets. Some are doctors, professors, salespersons in shoe stores, tax attorneys, shipping clerks, or teenagers who are just learning the basics. Others are professionals who have appeared on *The Tonight Show* or in traveling circuses. They all share a love of magic, and they willingly sit and watch each other's routines because they know how valuable it is to keep practicing.

2. Write a paragraph of at least five sentences in which at least four sentences have more than one independent clause. Underline the independent clauses you use. You may want to write about magicians or magic.

24b. Dependent Clauses

1. In the following paragraph, underline the dependent clauses and identify them as adjective or adverb clauses.

   There are an estimated 50,000 magicians in America. Most are amateurs who enjoy magic as a hobby. These amateurs often have elaborate equipment although their only audience is usually their friends and relatives. Some, however, specialize in the small card, coin, and rope tricks that are always popular. Purists call this intimate "close-up" magic the only real magic because it relies so heavily on a person's manual dexterity. Because people seem to prefer to be fooled face-to-face, this close-up magic is also offered by professional magicians who perform at birthday parties and trade shows. A psychologist who is also a magician says that when something is done under people's noses, it's more magical. It's much more elusive. The spectacular effects of magic done on television don't seem to impress people quite so much. Whatever the cause may be, amateur magicians will keep buying those sponge balls, decks of cards, special coins, and paper flowers.

2. Write a paragraph of at least five sentences in which three or more sentences have dependent clauses. Underline the dependent clauses. You may want to write about magic or magicians.

Chapter 25
Essential and Nonessential Clauses and Phrases

1. On the lines below, rewrite the following paragraph so that all the nonessential clauses and phrases are taken out of the sentence and put in a new sentence following the one they are in now. Identify the essential clauses by underlining them.

   Music teachers in the elementary schools who want to teach classical music to their students now have a lot of materials to help them. The materials, which many students say are excellent, have been developed by a group of educators who want to introduce classical music appreciation programs in elementary schools. These programs, incorporating listening skills and information about themes, styles, and forms of music, are also being used to coach students for contests. Students seem to enjoy these contests, a new form of competition in many schools. Contestants are often asked to identify the main themes of familiar works such as *Nutcracker Suite,* the specific instruments being played, and the composers. Harder questions, ones that stump even the more experienced teams, are those that are asked about music the students have not studied in school. Not all the music that is studied in the music appreciation classes is classical. Sometimes jazz and non-Western music is added, though students also seem to enjoy studying classical music that they have heard on television commercials. Music teachers who have tried these programs are pleased with the variety of types of music students come to like and the quantity of music they come to know. These music appreciation programs, begun on a small scale in a few states, are expanding rapidly across the nation.

   _____

   _____

   _____

   _____

   _____

   _____

   _____

2. Write a paragraph of at least five sentences in which at least three sentences have an essential clause and at least two sentences have nonessential clauses. Underline the nonessential clauses. You may want to write about studying music in school.

Chapter 26 Sentences

1. Identify each of the following sentences according to the letter which describes it.

    a. simple sentence (one independent clause)

    b. compound sentence (two or more independent clauses)

    c. complex sentence (at least one independent clause and at least one dependent clause)

    d. compound-complex sentence (at least two independent clauses and at least one dependent clause)

    e. incomplete sentence (does not have at least one independent clause)

___ 1. In today's society many working people who are in their thirties or forties face the problem of providing care for both their children and elderly parents while they are at work.

___ 2. Since children have less contact with older people than in previous generations, a growing number of programs are being designed to bring children and older people together.

___ 3. Which will benefit both groups.

___ 4. In some programs there are centers for child care next to centers for the elderly, and there are plenty of opportunities for visiting between the centers.

___ 5. Including shared activities such as cooking and birthday parties and informal get-togethers for storytelling hours.

___ 6. Psychologists point out that this kind of contact, even between old people and youngsters who are not related, can fill a void in young children's lives when they do not have a grandparent living near by.

___ 7. In addition, companionship with children keeps older people from feeling isolated and lonely.

___ 8. There is also a societal need for older people to transmit life experiences to younger generations.

___ 9. Something that is becoming increasingly less frequent in our mobile society where children move away from their parents and raise families on their own.

___ 10. Programs to link children and the elderly have been so successful that similar programs have also sprung up that link retirees with at-risk teenagers, and these programs are providing great benefits to the teenagers.

___ 11. A group based in Washington, D.C., which is a coalition of national organizations representing old and young constituents, works to push goals of mutual interest to the two age groups.

2. Write sentences illustrating each of the following types of sentences.

   a. compound sentence

   b. simple sentence

   c. incomplete sentence

   d. complex sentence

   e. compound-complex sentence

PART FOUR
PUNCTUATION

Chapter 27 Commas

27a. Commas in Compound Sentences

1. The following paragraph contains some compound sentences that need commas. Add commas where they are needed, and underline each comma you have added.

   In 1977 the flight of the little airplane, the Gossamer Condor, did not look very impressive but it was indeed a historic flight. With wings of foam, balsa wood, and Mylar, the plane designed by Paul MacCready floated slowly and gracefully over the San Joaquin Valley and covered a mile or so in about eight minutes. What made it so historic was that the pilot was pedaling. The Gossamer Condor was only the first of MacCready's pedal-powered planes and two years later the plane's successor, the Gossamer Albatross, crossed the English Channel. Some people say that MacCready is really the brains behind these inventions but others feel that he receives undue credit for the work that others on the development teams do. However, MacCready was the first to use his observations of how birds fly so his supporters feel that he is the genius who made human-powered flight possible. Other inventors were taking the conventional approach of trying to reduce drag as much as possible because they thought this approach would be the answer. The approaches of other inventors were to streamline their aircraft or they tried to incorporate ways to increase the horsepower. Only MacCready applied the principle of vastly increasing the wing area and used materials to keep the overall weight down. The result was an aircraft that needed only the power output of a good bicyclist and the Gossamer Condor now has a place of honor next to the *Spirit of St. Louis* in the Smithsonian Institute's National Air and Space Museum.

2. Write a paragraph with at least five compound sentences that are punctuated correctly with commas. You may want to write about the development of the airplane.

   _____

   _____

27b. Commas After Introductory Words, Phrases, and Clauses

1. The following paragraph contains some sentences with introductory words, phrases, and clauses that need commas. Add commas where they are needed, and underline each comma you have added.

      Having won prizes with his first human-powered plane, Paul MacCready went on to build a faster and more powerful pedal-powered plane, the Gossamer Albatross. In less than two years after his first success, MacCready's second pedal-powered plane departed from Folkestone, England, in June, 1979, bound for France. Expecting the flight to take about two hours, MacCready allotted just enough water for the pilot to drink. The flight team who prepared the plane and assisted the pilot on the ground waited several weeks for the kind of calm weather that was needed. Consequently, the pilot took off, expecting to reach France before his endurance and the water gave out. But a head wind blew up soon after the pilot was aloft. An hour and a half later, he was only two-thirds of the way to France, and his legs were cramping from all the pedaling. Because everyone was sure they had to give up the attempt, the flight team was ready to hook a towline to the craft that would haul it ashore. Tired and about to give up, the pilot knew he had to gain altitude to get hooked to the towline. As he climbed, he found less wind and was able to press on. Almost three hours later, the pilot touched down at Cape Gris-Nez, in France, a minute short of his theoretical exhaustion point. The Gossamer Albatross had crossed the English Channel, powered only by the pilot.

2. Write a paragraph with at least five sentences that have introductory words, phrases, and clauses correctly punctuated with commas. You may want to write about a strenuous sport or competiton you were in.

_____

_____

_____

_____

_____

_____

_____

_____

_____

_____

27c. Commas with Essential and Nonessential Words, Phrases, and Clauses

1. The following paragraph contains some sentences with nonessential words, phrases, and clauses that need commas. Add commas where they are needed, and underline the commas you have added.

After designing human-powered planes, Paul MacCready a prize-winning inventor went on to design a solar-powered plane. MacCready however realized that solar cells as an energy source for planes do not make any practical sense. But MacCready who had long sympathized with environmental concerns hoped to demonstrate that solar power has an important part in the world's energy future. Those who see solar energy as merely a minor source of energy for the future downplay the importance of such demonstrations. Others think solar power has simply not been adequately developed for practical use. The solar-powered plane that MacCready designed flew from Paris to the coast of England in 1981 cruising at 441 mph at an altitude of 11,000 feet. The plane called the Solar Challenger provided the stepping stone to MacCready's next flying machine the Sunraycer a solar-powered car.

2. Write a paragraph with at least five sentences that contain essential and nonessential words, phrases, and clauses correctly punctuated with commas. You may want to write about the future of solar energy.

_____

_____

_____

_____

_____

_____

_____

_____

_____

_____

_____

27d-h. Commas in Series and Lists; Commas with Adjectives; Commas with Dates, Addresses, Geographical Names, and Numbers; Other Uses for Commas; Unnecessary Commas

The following paragraph contains some sentences with correct comma usage and with some comma errors. Add commas where they are needed, omit unneeded or incorrect commas, and underline each change you make.

      The Sunraycer which is a solar-powered lightweight car, was built to compete in the 1987 Race Across Australia. Designed by Paul MacCready the car won the race from Darwin to Adelaide, and is now in the Smithsonian Institute's National Museum of American History. With a total weight of 365 pounds the car has a power output of about 1.8 horsepower at noon on a bright day and it gets the electric power equivalent of 500 miles to the gallon. The Sunraycer which presently holds the solar-powered speed record of 48.7 mph averaged a little over 40 mph for much of the race. The car is so light, that when it made turns during testing it often seemed in danger of blowing over. The engineers who worked on the Sunraycer, ended up putting two little ears on the top. "We're not sure why they work" said one engineer "though they seem to help." In some ways the Sunraycer is not a prototype of electric cars for commercial use, because the Sunraycer has bicycle-thin wheels a driver's seat that requires the driver to lie flat, and very weak acceleration. However many of Sunraycer's features were carried over into the electric car developed by General Motors. Like the Sunraycer the GM car uses alternating current, and can therefore get better performance. If electricpowered cars become widely popular in the future Paul MacCready will feel that he had some small part in saving the environment.

27a-h Review of Comma Usage

Commas are used correctly throughout the following paragraph. On the lines below, explain why the commas in each sentence are necessary.

(1)At the turn of the century Howelton was a well-established community of family-operated small farms at the top of Little Sand Mountain in Etowah County, Alabama, between Attalla and Walnut Grove. (2)In this community Thomas Lee and Mollie Benson grew up. (3)He was born in 1888, the son of M.O. and Alice Lee, and Mollie was born to W.J. and Fanny Benson in 1891. (4)Both families were highly respected, hard-working members of the community. (5)Both Thomas and Mollie completed the seven grades available at the community school, which meant they were prepared to take their places in the community and to take on the responsibility of marriage and family. (6)Though a few would move to one of the nearby towns to get a job in public works, most young people would at that point become farmers. (7)Having been sweethearts for years, Mollie and Thomas were married on Sunday, January 31, 1909, when he was 21 and she was 18. (8)In that community at that time there were very few big weddings. (9)Usually the couple informed the minister that they wished to be married at the end of the morning service, and they had a simple ceremony with friends and family in the congregation. (10)Thomas and Mollie's wedding, however, was more memorable than most, as the church burned to the ground that afternoon!

1. _____
   _____

2. _____
   _____

3. _____
   _____

4. _____
_____

5. _____
_____

6. _____
_____

7. _____
_____

8. _____
_____

9. _____
_____

10. _____
_____

Chapter 28 Apostrophes

1. The following paragraph contains sentences with some apostrophes incorrectly used to show possession, mark contractions, and indicate plurals, as well as some missing apostrophes. Add apostrophes where they are needed (even if they are optional), omit those that are unneeded or incorrect, and underline each change you make.

   In the game of baseball, batting slump's are one of a players worst nightmares. When they are doing well, player's attribute their successes to mysterious minor occurrences around them that then become habits' the players keep up. After a game in which one baseball player who was wearing an old helmet with it's side dented hit two triple's, a home run, and a single, the player continued to wear that helmet for the rest of the season. Warding off evil spirits through superstitions is another thing baseball player's do. One player always wears the green T-shirt from his university under his teams jersey. Another wont wear a jersey with any 6s in his players number. Batting coaches spend hours watching videotapes with slumping players, trying to find whats causing the problem. They examine the players batting stance or swing, but this does'nt always provide useful clues. Some slumps happen when batters begin to worry too much about their misses and about everyone elses successes. But one coach thinks otherwise. He notes that some players' start making adjustments when theyve hit a double and want to hit farther or when a certain unusual pitch connected well with their bats. Players in their 30s' complain of a different kind of slump. One bad day, says one over-30 player, may mean he is losing it, that his age has begun to take it's toll. A hitter who is hot, on the other hand, tends to get possessive about a special bat and uses it until it breaks or its' cracks begin to show. When that bat goes, the player sometimes loses confidence until he connects with a new bat that brings him a few homerun's. A batters life is'nt as easy as some people think it is.

2. Write a paragraph with at least five sentences that use apostrophes correctly. You may want to write about the performance slumps of professional athletes or teams or about low periods or slumps in your life.

   _____
   _____
   _____
   _____

Chapter 29 Semicolons

1. The following paragraph contains some sentences that need semicolons and some sentences with incorrectly used semicolons. Add semicolons or change punctuation marks to semicolons where they are needed, delete those that are incorrect, and add correct punctuation, if needed. Underline each change you make.

   Junk mail used to be confined to print on paper, now it is appearing in people's mailboxes on videocassettes. Companies in the direct-mail business are now marketing inexpensive cardboard videocassettes that can carry a variety of messages; such as audiovisual advertisements, promotional premiums, and educational or training aids. Some companies are switching to this form of direct-mail advertising because it is relatively cheap, in addition, it presents messages more vividly on television screens than print advertising can on paper. Informational videos can be sent to prospective customers, and advertisers can use their product in the video. Says the spokesperson for one cereal company, "We are interested in promoting good nutritional habits;" as might be expected, the balanced diet they will be picturing in the video will include their cereal. Printed instructions on merchandise are often confusing, consequently, some manufacturers are also switching to these disposable videocassettes for the instructional packets they include with their merchandise. The cardboard videocassettes are relatively cheap to manufacture and cheap to mail. They are certainly more convenient than the promotional packets sent by companies that have relied on enclosing sample packets of toothpaste, aspirin, or cereal, mail advertisers who send bulky envelopes of coupons, and companies who want to entice customers with big brochures of vacation places, hotels, and tours. The disposable videocassette is certainly going to grow in popularity as an advertising medium.

2. Write a paragraph with at least five sentences using semicolons in compound sentences and lists. You may want to write about junk mail.

   _____

   _____

   _____

   _____

   _____

Chapter 30 Colons

1. The following paragraph contains some sentences that need colons and some sentences with incorrectly used colons. Add colons or change punctuation marks to colons where they are needed, delete those that are incorrect, and add correct punctuation, if needed. Underline each change you make.

   Evaluating college teachers is a complex, often difficult task. Now a new method devised in the Midwest is being tried out across the nation the teaching portfolio. Administrators and faculty members see the portfolio, a collection of materials documenting classroom performance, as a way to emphasize teaching as a major priority. Too often, professors who are being evaluated document primarily their research and scholarly activities, such as: published books and article, lists of grants, and conference presentations. Says one faculty member at a large Eastern university, "When it comes to teaching, most teachers only have student evaluations and coffee-room conversations about what they do in class and how their students are learning": this kind of evidence is not very thorough or sufficient to determine pay raises, tenure, and promotion. As a result, many faculty members are judged and rewarded on the basis of their research and publications. Their teaching performance, whether outstanding or mediocre, is largely ignored. The portfolio can be a better measure of what faculty really do as teachers, for the portfolio can consist of the following items: a statement of the person's teaching philosophy, a list of the courses that person has taught, a representative syllabus, statements by fellow faculty members who have observed this person, and course materials that the person has prepared. At some schools where portfolios are already in use, the portfolios also include: unsolicited letters from former students, teaching awards, and even videotapes of the candidate teaching a class. Some faculty members who have prepared these portfolios say that there is an added advantage, the portfolio collection can lead to self-improvement. This can happen when the process of creating the collection causes the person to think seriously about teaching goals, strategies, and results.

2. Write sentences using colons in each of the patterns below.

   a. Colon to announce an element at the end of the sentence

   _____

   _____

b. Colon to separate independent clauses

c. Colon to announce a long quotation

d. Colon in a salutation

e. Colon between elements

f. Colon with quotation marks

Chapter 31 Quotation Marks

1. The following paragraph contains some sentences with correct and incorrect use of quotation marks. Revise and add quotation marks, if needed. Write your changes in above the lines.

    "Why am I fatter than my sister-in-law? I eat less, complained a woman being studied by a team of researchers. She explained that she "repeatedly went on diets when her sister-in-law didn't." But the woman continues to weigh more. Researchers are finding out that heredity, in addition to "lifestyle,î"exerts a strong influence on people's weight. By studying identical and fraternal twins, research teams are finding that brothers and sisters end up with similar body weights whether or not they are raised in different families. In the "Journal of Genetics," Dr. Albert Skinnerd writes, "When the biological parents are fat, there is an 80 percent chance that their children will also be overweight". (234). "Does this mean that my brother and I are doomed to be fat," asked one overweight twin in the study? Since some sets of twins tend to transform extra calories into fat while other sets of twins tend to convert extra calories into muscle, one scientist concluded that "genes do seem to have something to do with the amount you gain when you overeat". Some unsuccessful dieters may be relieved to know that their failed diets aren't a matter of failed "willpower." It is really a matter of metabolism", reports another doctor doing research in this field. But that does not mean that low-fat diets and exercise should be given up. Quit is not a word in my vocabulary", says one constant dieter who manages to maintain a reasonable weight by means of careful eating and plenty of exercise, despite a tendency to be overweight.

2. Write a paragraph of at least five sentences using at least five sets of quotation marks for direct quotations, minor titles and parts of wholes, words used as words, and other uses of quotation marks. You may want to write about dieting.

    _____

    _____

    _____

    _____

    _____

3. Write a dialogue based on this situation: You and two classmates are discussing an upcoming exam in your American literature class. Include at least six separate speeches (two by each speaker, for example) and include titles of a short story, an essay, or a sermon; a poem; and a novel that may be covered on the test.

Chapters 32, 33, 34
Hyphens, End Punctuation, and Other Punctuation

1. The following paragraph needs one or more hyphens, dashes, slashes, parentheses, and brackets. Add those that are needed, revise any that are not used correctly, and underline all changes that you make.

   Which oils are good for us to eat? A study of thirty nine participants on a reduced fat diet looked at the benefits of consuming olive oil and [or] corn oil. Which is more beneficial in influencing high/density lipoprotein levels corn oil or olive oil? This is an important question because high-density lipoprotein [HDL] is considered a beneficial form of cholesterol that helps remove the more dangerous low-density lipoproteins from the body. In a study published in an article, "Two Healthy Oils for Human Consumption" [*Diet and Health News* 14 (1991): 22636], researchers report that the participants first spent twelve weeks on a diet that included olive-oil and then another twelve weeks on a corn oil diet. The results (which were also announced on television newscasts) indicated that neither diet resulted in lower-levels of HDL. This indicates that a diet of olive oil and (or) corn oil can accompany a reduced—fat diet. For those who self-select the oils they use in their diet, the choice is probably a matter of taste-or cost.

2. The following paragraph contains some sentences with incorrect use of periods, question marks, exclamation points, and ellipses. Add or change these punctuation marks where they are needed, delete those that are incorrect, and add correct punctuation, if needed. Underline each change you make.

   "Do you see that green area to the left of the river we are flying over"? said the pilot to the passengers as the commercial 747 jet flew over southern Texas. Continued the pilot, "That's my grandfather's ranch! I often visited there as a kid!" Public address systems in commercial planes are now being used by pilots to enliven their passengers' flights. Some pilots are opposed to this practice because they see it as a distraction. Says one seasoned veteran, "Our task is to fly the plane, not amuse the passengers". But others disagree. Interesting, informative comments can put nervous passengers more at ease and can shorten a long flight. For those pilots interested in making such publicaddress announcements, there is now a book put together by an ex-pilot which pinpoints more than 1,200 historical and little-known places of interest on a collection of highway maps of the U.S. The maps are overlaid with the flight paths used by commercial pilots. Are you flying between El Paso, Texas, and Las Vegas, New Mexico. If so, then look for the site of the Berringer Crater, where a meteor hit with such force 22,000 years ago that it killed all animal and plant life within 100 miles! Thousands of these maps have been sold, with more frequent fliers than pilots doing the buying! 'We are learning…an awful lot about territory we thought we knew"! says one frequent flier who takes her book with her on every flight. Another customer reports, "I bought one for my uncle who hates airplanes, and he now actually enjoys his flights". The book is obviously a success! Is the author correct when he says, "Soon, there will be such books in the pocket of every seat in every commercial flight?

3. Write a paragraph of at least five sentences correctly using one hyphen, one dash, one slash, one ellipsis, one set of parentheses, and one set of brackets. Also use three different kinds of end punctuation correctly. You may want to write about commercial flights.

Chapters 27-34

Copy a paragraph of eight to twelve sentences from a paper you have recently written or are now working on. Number and correctly punctuate every sentence. On the lines below the paragraph, write the number and provide a brief explanation for each punctuation mark in every sentence.

Explanation:

PART FIVE
MECHANICS AND SPELLING

Chapters 35, 36, 37
Capitals, Abbreviations, and Numbers

1. Revise the following paragraph for correct use of capitals and abbreviations, as well as the correct way to write numbers. Underline all changes that you make.

      One of Boston's most popular folk heroes is Paul Revere, whose Midnight ride was made famous in Longfellow's poem, "Paul Revere's Ride," in *Tales Of A Wayside Inn.* Not many tourists realize that many of the places mentioned in that poem are still part of Boston's Twentieth-Century landscape, along what is commonly called the Freedom trail. This well-marked Trail is actually a 2-mile line in red paint, bricks, or steps with 21 historic buildings, sites, and monuments documenting Boston's contribution to American history. 2 miles isn't long, but it usually takes tourists who are thorough 2 or 3 days to see everything on the trail. Most begin at the Boston common, the oldest public park in the U.S. Here colonists used the public land for grazing animals, training the militia, and staging Public executions. Immediately next to the Common is America's first public botanical garden, the Boston Public garden, famous partly because it was the setting for the well-known children's book, *Make Way For Ducklings.* Just across the Street, on Beacon Street, is another of Boston's famous sites, the Old State House, completed in Seventeen Ninety-Eight. Its cornerstone was laid by Samuel Adams in 1793. 1 block East on Beacon street and then 1 block south on Park street is the historic Park street Church on the corner of Tremont. Here the song "America" was first sung on July 4, 1831. The next stop on the Freedom Trail is the Granary burying Ground, a Cemetery on Tremont Street. Some of the greatest U.S. patriots, including Paul Revere, Samuel Adams, John Hancock, James Otis, etc, are buried here. Also along Tremont street, at the corner of School street, is King's Chapel, the first anglican church in Boston. After the revolution, it became the first unitarian church in the country, and its cemetery contains the graves of many other early notables, including William Dawes, Jr. and Gov John Winthrop. (it was Gov. Winthrop who headed the group that founded the Boston colony.) Also on School Street is the first public school in america, which later became the Boston public latin School. Rev Cotton Mather, Ralph Waldo Emerson, John Hancock, Benjamin Franklin, et al, were alumni of this School. These are only a few of the many sights along the Freedom Trail. Longfellow's poem says, "hardly a man is now alive who remembers that famous night and ride...." True, but the Freedom Trail helps visitors realize many of the events that led to our Nation's independence.

2. Write a paragraph with at least five sentences correctly using capitals and numbers. Also include abbreviations of numbers, measurements, and dates. You may want to write about history.

Chapter 38 Underlining/Italics

1. Listed here are titles and words that either need underlines (or italics) or quotation marks. In the right-hand column, write a "U" if the titles or words need an underline (or italics) or "Q" if the titles or words need quotation marks.

   a. Newsweek (magazine) — U

   b. Gone with the Wind (movie) — _____

   c. My Next Adventure (book chapter) — _____

   d. Buffy the Vampire Slayer (TV series) — _____

   e. Moby Dick (book) — _____

   f. To His Coy Mistress (short poem) — _____

   g. The Case of the Lost Diamond (TV episode) — _____

   h. The Barber of Seville (opera) — _____

   i. Paradise Lost (long poem) — _____

   j. Copenhagen's Museums (magazine article) — _____

   k. Time (news magazine) — _____

   l. Sweet Birdie (song) — _____

   m. Civil Rights Revisited (pamphlet) — _____

   n. The Economics of Solar Energy (magazine article) — _____

   o. C (letter of the alphabet) — _____

   p. Jurassic Park (book) — _____

   q. Exxon Valdez (ship) — _____

   r. dejure (Latin phrase) — _____

s. My Nervous Childhood (essay) _____

   t. Washington Post (newspaper) _____

2. Write a paragraph of at least five sentences in which you correctly use underlining at least five times. You may want to write about books, magazines, articles, television programs, and movies that you enjoy.

Chapter 39 Spelling

39a. Proofreading

Proofread the following paragraph for spelling errors, typos, and omitted words. Use the dictionary if needed, and underline all the revisions you make.

Inventors of gadgets for automobiles havn't always been successful with there inventions. But we can see from some these inventions that people have been looking for ways to make cars more functional, better looking, and more fun to drive, for example, we now have elegent and sophisticated ways tohere music in ou cars, but some of the earlier ways to add music to driving seem a bit odd now. In the 1920's, Daniel Young recieved a patent for an organ he invented for use in automobiles. He built organ keyboards that could be attached to back of the front seat so that people ridding in the back could play the organ to entertain themselfs. This may have been a good idea, exept for one thing. The roads of that time, unfortunatly, were so bumpy and uneven that the sounds producted by the organ when the car was moving where anythhing but beautiful. Another terriffic idea that didn't make it was Leander Pelton's patent for a car that could be parked by standing it on end. Instead of a back bumpper, he buildt a vertical platform with rolers attached. When parking the Vertical-Park Car, the driver needed to tip the car back onto platform. Than he could just shove the car into any approrpiately sized space. To preform this task, however, was a big dificult as Pelton never quite explained how the car was to be tiped from horzontal to vertical and back down again. A diffrent problem was that Pelton didn't porvide any way to keep gasoline, water, and oil from spilling once the car was up on its parkking rollers. But the government gave him a patent; he simply couldn't get anyone to manifacture his VerticalPark car. Another invention that never made it was designed by Joseph Grant in 1926—an autombile wash machine. The machin didn't wash cars, but it supposssedly washed clothes. Grant's invention consistted of a tub and paddles that bolted to the car's runing boards. when the tub was filled with water, soap, and drity clothes, the bouncing of the car over rought roads provided all the power and agitatition necessary to clean a load of dirty clothes. For realy dirty load, an extra twenty miles or so of driving was reccommended.

39c. Some Spelling Guidelines

1. **IE/EI:** Using the rules for IE/EI spellings (and the exceptions), underline the correctly spelled words in these sentences.

   a. Does he really (beleive, believe) he will win the lottery?

   b. The soybean (yeild, yield) this year will be high.

   c. She knows how to (seize, sieze) the day.

   d. The (height, hieght) of that tree is unusual.

   e. The company is working hard to establish more (foreign, foriegn) markets.

   f. My (weird, wierd) brother is always trying some crazy new herbal diet.

   g. What is your (field, feild) of study?

   h. No more than (eight, ieght) people can fit in that van.

   i. He tried to (deceive, decieve) himself into thinking he was able to do that.

   j. She approached a (financeir, financier) to help her start a new business.

   k. We called off the party when (neither, niether) of the guests of honor could come to it.

   l. During the rainstorm, the (ceiling, cieling) leaked badly.

   m. When is your (neice, niece) coming to visit?

   n. Our new (neighbor, nieghbor) seems like a pleasant person.

   o. The nurse tried to find a (vein, vien).

2. **Doubling Consonants:** Underline the correctly spelled word in these sentences.

   a. I have to be quiet when my sister is (naping, napping).

   b. It was hard to keep my (footing, footting) on that slippery slope.

   c. Was that the movie that Paul Newman (starred, stared) in?

   d. As she listened to the music, she kept (tapping, taping) her fingers on the desk.

   e. To whom were you (writting, writing) that long letter last night?

   f. I hope she (referred, refered) me to a good dentist

   g. When I stubbed my toe, I (hopped, hoped) around moaning and complaining.

   h. It was a strange (occurrence, occurence) to see the sky lit up like that.

i. I came in at the (begining, beginning) of the movie.

j. They really (benefited, benefitted) from that study session before the exam.

k. My father hates to go (shoping, shopping) when he's on vacation.

l. The advertisement (omitted, omited) any mention of the price.

3. **Doubling Consonants:** Write sentences using the -ed or -ing forms of the words listed here in sentences.

   a. shop

   b. hope

   c. tape

   d. star

   e. write

   f. top

   g. slip

h. omit

i. prefer

j. refer

k. benefit

l. occur

4. **Prefixes and Suffixes:** Correctly add the prefixes or suffixes to the words listed here, and then write a sentence using the word and its prefix or suffix. Consult a dictionary if needed to check on the spelling or the meaning of the word that is formed by adding the prefix or suffix.

   a. actual+ly

   b. ante + cedent

   c. anti + biotic

d. auto + biography

e. bene + diction

f. bi + partisan

g. desire + able

h. de + value

i. dis + agree

j. im + migrate

k. inter + mission

l. mis+ inform

m. mis + spell

n. notice + able

o. pre+form

p. per+form

q. picnic + ing

r. pro+nounce

s. real+ly

t. true + ly

5. **Y to I:** Correctly add the suffixes given to the words listed here, and then write a sentence using the newly formed word.

   a. ready + ed

   b. lonely + ness

   c. carry + ed

   d. carry + ing

   e. bounty + ful

   f. greasy + ness

   g. play + ed

   h. employ + ed

i. petty + ness

_____

_____

j. forty + eth

_____

_____

39d. Plurals

For each of the following words write a sentence using the plural form of the word.

1. apology

_____

_____

2. ceremony

_____

_____

3. box

_____

_____

4. city

_____

_____

5. sister-in-law

_____

_____

6. church

_____

_____

7. videocassette recorder

_____

_____

8. radio

_____

_____

9. basis

_____

_____

10. attorney

_____

_____

39e. Sound-Alike Words (homonyms)

1. Underline the correct word in these sentences.

    a. Everyone came to the concert (accept, except) Tiana.

    b. Does that store (accept, except) personal checks?

    c. I think the horror movie he saw (affected, effected) his sleep.

    d. The horror movie had no (affect, effect) on him whatsoever.

    e. When her friend is visiting (hear, here), she never gets any studying done.

    f. (It's, Its) a cool, rainy day today.

    g. I wonder when (itís, its) going to get better.

    h. What is (it's, its) temperature?

    i. Last week (passed, past) so quickly, it seems.

    j. She is no bigger (then, than) I am.

    k. Leah stayed until her friend joined her, but (then, than) she left the party.

    l. What did they do (than, then)?

    m. All the speakers (there, their) vowed (there, their) loyalty to the union.

    n. I gave my red sweater (to, too) my sister since I have (to, too) many.

    o. They (were, where) on the right road, but they took the wrong turn.

    p. I never know (whose, who's) right in those arguments.

    q. Kristen knows (your, you're) right, but she will never say it to (your, you're) face.

    r. Could you give me some (advice, advise) about (buying, bying) a used car?

    s. That is going to be the (sight, site, cite) for the new shopping mall.

    t. She was (quite, quiet, quit) sure that she had taken her wallet out of her purse.

    u. You have to keep the camera (stationary, stationery) when you take the picture.

    v. Everyone had (all ready, already) left when the music finally began.

    w. I guess it's (alright, all right) to charge that sweater to my mother's account.

    x. Perhaps you can use (anyone, any one) of the golf balls lying there.

    y. The kit to build that model comes (altogether, all together) with the tools you will need.

2. Use the following words in sentences. Be sure that they are spelled correctly

    a. accept

    _____

    _____

    b. except

    _____

    _____

c. affect

d. effect

e. here

f. hear

g. its

h. it's

i. passed

j. past

k. than

l. then

m. they're

n. their

o. there

p. too

q. to

r. were

s. we're

t. where

u. who's

v. whose

w. your

x. you're

y. advice

z. advise

aa. site

bb. cite

cc. sight

dd. dessert

ee. desert

PART SIX
STYLE AND WORD CHOICE

Chapter 40 Sexist Language

1. The following paragraph has some sexist language that can be revised. On the lines below, rewrite the paragraph so that sexist language has been eliminated, and underline all the changes that you make. You may need to make other changes in various sentences as well.

In many suburban housing developments built during recent decades there are homeowners associations that enforce housing codes on all the homeowners. The average owner in such a suburb may think that he is free to paint his house whatever color he likes or park any kind of car in his driveway, but that is not the case. Homeowners associations often have a lawyer who spends his days enforcing the laws enacted by these associations. The laws see to it that all the members abide by the group's standards of good taste. No plastic flamingos are allowed on the lawn, and every house painter who works in the suburb knows that he cannot use certain colors for house trim, such as bright pink or a gaudy yellow, because he has to follow community guidelines. When houses are built in new developments, there are usually restrictive covenants that force the buyer to join the association, whether he likes it or not. Even when someone challenges the laws, he usually loses as the covenants are legally binding. It all starts with the builder because when a builder builds, he wants to make certain that the land value for the community stays high so that he can continue to sell his houses at a good price. The builder often starts off as the chairman of the homeowners association so that he can guide the formation of the rules and regulations. In one wealthy Florida community there are even regulations for local government and civil servants, including dress codes for policemen, taxi drivers, and mailmen. The only challenges that have gotten through the courts are those that show some regulation discriminates on the basis of race, religion, sex, or other characteristics of the homeowner.

_____

_____

_____

2. Write a paragraph about a job you have had in college or in high school or about the career you hope to pursue when you have graduated from college. Write the paragraph in third person, not in first person, and include a description of the responsibilities and expectations of persons in that job. Avoid sexist language in your description.

Chapter 41 Unnecessary Words

41a. Conciseness

The following paragraph contains many words and phrases that are unnecessary. Rewrite the paragraph on the lines below so that it is more concise.

It is definitely true that, as we all know, the topic of telling time is one that is of great importance to us all. This paragraph will discuss here the topic of time in our daily lives and how the concept of time has changed over time through the ages. Time plays an important role in people's lives because it is the essential measure against which other measurements of great and necessary importance were made. For example, we know that we can see that we measure the heart rates of our bodies in terms of time, and we know that we measure how fast our cars travel in terms of time. We organize our days and nights into whole segments of time, such as days, hours, minutes, and seconds, and from the very earliest beginnings of civilization people have counted the passage of time in terms of counting sunrises and sunsets or the movements of the moon as well as the movements of the sun in the sky overhead. For many centuries there was speculation about the nature of time that was mostly a philosophical discussion of how people perceive time and experience the passage of time. But in the twenty-first century science of today, since the work done by Albert Einstein, physicists have now come to realize that time is a definite dimension of the physical universe. Time is a measure of motion in space, not just some philosophical or theoretical thing that exists in people's minds as a concept they think about. The work of Albert Einstein, who was a physicist, also showed that time is not an absolute thing and that there is no such thing as a unique absolute time. It is known that people used to think that any event measured in time would be seen to take the same amount of time. To give an example of what this means, people used to think that two good clocks which are in good, accurate working order would agree on the time interval that it takes between two events. But history books tell us that the discovery that the speed of light appeared the same to every observer, no matter how he or she was moving, led to the theory of relativity. Now we know that time is seen by everyone relative to the observer who measures it. Each observer who observes time can have his or her own measure of time as recorded by the clock that he or she carries. Clocks that are carried by different observers do not necessarily have to agree. This view that observers do not have to agree on the time interval is a very different view of time from the older view of time that it is an absolute thing. However, as shown here, even with this notion that time is not an absolute thing, we still in our lives today use time as a means of measurement.

41b. Clichés

Listed here are some word groups that many people regard as clichés. Try to think of a different way to express the same idea by writing a sentence using fresher language instead of the cliché.

1. white as snow

2. rain or shine

3. beat around the bush

4. easier said than done

5. big as a house

6. work your fingers to the bone

7. live happily ever after

8. first and foremost

9. to be turned on by

10. climb the ladder of success

41 a & b Conciseness and Clichés

This paragraph is characterized by wordiness and use of clichés. Rewrite it on the lines below so that it is concise and precise.

Many people in this day and time are completely devoted to their pets that they think they cannot live without. People particularly seem to feel this way about their dogs, although probably some people feel that way about cats too. Whether the pet is a gigantic mastiff that looks wise as an owl or a big-eared Chihuahua that yaps all the time at everyone and everything it sees, all owners consider their precious pets to be their pride and joy. Reasonable, rational, thinking pet owners seem to be few and far between. Some people dress their pets up in expensive clothes so they look fit to kill. They may buy pet clothes on the Internet or in fancy pet stores in up-scale malls in big cities. They may even buy beds for their pets that cost an arm and a leg and send the pets to doggie day care when they have to go to work. They would even probably take their pets to work with them if they were allowed to do that at the place where they work. People who don't own pets may consider their friends who do own pets to be crazy as a bed bug because they spend so much money, time, and energy on their pets. But true pet lovers just can't seem to face the music without their pets; they feel more calm, cool, and collected when their pets are around, and they are willing to spend whatever they feel is necessary to keep their pets content and happy.

41c. Pretentious Language

1. Listed here are some sentences that use what many people regard as pretentious language. Try to think of a different way to express the same idea by writing a sentence using plain English instead of the pretentious language. You may want to use a dictionary to determine the meaning of some words.

   a. I unquestionably ascertained that the oration I planned to deliver to the assemblage of students required considerable revision.

   b. I was rendered mute by the import of the situation.

   c. Nathan is someone who is noted for his singularity.

   d. I shall slumber anon and awaken refreshed to meet the challenges of the pristine day.

   e. The lucidity of the discourse was the source of significant veneration.

2. The language in the following paragraph is pretentious and wordy and thus inappropriate for most audiences. Rewrite the paragraph so that the language is appropriate for the target audience of college students.

As the twenty-first century moves inexorably forward in America, institutions of higher learning are endeavoring to appeal to diverse student populations and provide unprecedented access to their resources by making education available in multitudinous formats. Scholars may pursue their studies in traditional classroom settings, or they may opt to complete an entire degree program without ever gracing a college campus with their presence by conducting all their coursework in the amorphous environment of the Internet. Courses that do not require the student's physical presence on campus are called online courses and are conducted in their entirety through institutional electronic facilities. The student who is contemplating attempting this medium of intellectual pursuit must carefully consider the personal requirements for success in the online environment. The prospective online scholar must be highly disciplined, with no propensity toward procrastination. He or she must possess an elevated sense of personal responsibility and an advanced reading level. The preponderance of evidence supports the contention that, contrary to the collective opinion of older adolescent computer literati, online pursuit of college credit is not facile and effortless. In point of fact, online classes will indubitably require a far greater investment of time and concentrated effort than traditional classes. Nevertheless, earning college credit through an online course will culminate in augmented computer literacy as well as discipline-specific erudition.

Chapter 42 Appropriate Words

42a-b. Standard English, Colloquialisms, Slang, and Regionalisms

Listed here are sentences that use some nonstandard words, colloquialisms, slang, and regionalisms. Try to find a different way to express the same idea by writing a sentence using standard words in place of the nonstandard and by using appropriate words in place of colloquialisms, slang, and regionalisms.

1. I ain't planning to take those kids to the circus.

2. Jason watched a sci-fi movie and then zonked out on the sofa.

3. Mark weren't planning to fight, but when Frank started to dis him, he decided he couldn't be a chicken.

4. I'm not uptight about the English exam, but I'm afraid I'm going to flunk the history exam.

5. Dad freaked out when he saw I had my tongue pierced.

42c. Levels of Formality

The following paragraph is written in a very informal tone. Revise it so that it is more formal and would be appropriate to print in a science news magazine intended for a general audience of reasonably well-educated readers.

Parents worried about their kids being grabbed by kidnappers have a new gimmick. Parents used to depend on taking photos with the old Polaroid or maybe getting a set of prints to stash in some computer file. But those methods aren't foolproof. Kids grow up and out of what they looked like in photos, and prints can be kind of hard to match. Now scientists bugged by not having a really foolproof way to nail a kid's identity have found a way that is looking good. Parents can get a sample of their baby's DNA material, which has great gobs of information about that particular kid. There's even a storage place set up for this in New Jersey. What's on file there, if someone wants it, is enough info to take a sample and match it against a hair, piece of skin, or even a couple of dots of blood. This is called DNA "fingerprinting," and it nails a kid's identity positively. What's really cool about this is that all the information from the genes stored at this place lasts as long as the person hangs around this planet—or longer. Grown-ups are pretty interested in getting themselves on file there too, especially those whose personal safety isn't high because of a dangerous job.

42d. Jargon and Technical Terms

Listed here are some sentences that use jargon and technical terms. Try to find a different way to express the same idea by writing a sentence using easily understandable language for a general audience. You may want to use a dictionary to find the meaning of unfamiliar words and terms.

1. My doctor told me that the antibiotic he prescribed for the inflammation of my gastrointestinal tract is counteracted by over-the-counter cold medications.

    _____
    _____

2. The learning facilitator of our pottery-making class doesn't receive enough monetary remuneration.

    _____
    _____

3. Our supervisor has asked that we revisit the problematic situation and provide input to its resolution.

    _____
    _____

4. Each candidate promised to implement no new revenue enhancements.

    _____
    _____

5. The scientist was studying the ferromagnetic properties of various Paleolithic artifacts.

    _____
    _____

42e. General and Specific Words

The following paragraph contains some underlined general terms which need to be replaced by more specific terms. The paragraph also contains some underlined specific terms which need to be replaced by more general terms. Revise the paragraph by replacing these terms with more appropriate ones. Write your revisions in above the lines.

People like driving up for fast food, and now so do their <u>animals</u>. A new fast-food industry has begun for drive-in dog food, and the menu is entirely for <u>spaniels and collies</u>. These new <u>things</u> offer treats to dogs with dog biscuits shaped to resemble food. The dog biscuits are made from foods that help keep dogs healthy. The biscuits are <u>flavored</u> so that dogs don't get tired of the same thing. Customers love the idea of <u>driving their</u>

trucks to a doggy drive-in after picking up their own stuff. So far the menu has been limited to dog biscuits, but some men will come up with new ideas for better bone-shaped biscuits for dogs.

42f. Concrete and Abstract Words

The following paragraph contains some abstract words that are underlined. Change those abstract terms to more concrete ones.

When hikers reach a stream, they often decide to cross where the route approaches the water. But this may not be the best place at which to cross. Water usually moves most swiftly at the narrowest part of the stream. So, hikers should instead look for another spot where the stream widens. Here the current often has less velocity and may be easier to walk through. When hikers are carrying a backpack, they should loosen the shoulder straps and hip belt before immersion so that they can toss off the pack if difficulties occur. Some hikers find that if they suddenly hit a depression in the streambed, the weight of the backpack can toss them off balance. Another aid to crossing a stream is a good hiking stick. It can serve as another leg, offering better balance when there are dangerous elements present. It is helpful to remove unneeded clothing before crossing a stream with a swift current because the water can drag against wet outerwear. The hiker should also take each step slowly and deliberately. The forward foot should be planted firmly before the rear foot is moved. Finally, the careful hiker never hurries across a stream.

42g. Denotation and Connotation

1. Each of the following words has either a positive or negative connotation. Think of an alternative term that has the opposite connotation.

EXAMPLE

devoted   obsessed

a. notorious   _____

b. assertive   _____

c. right-wing   _____

d. orderly, well organized   _____

e. garbage collector   _____

f. pre-owned _____

g. dirt _____

h. fragrance _____

i. economical _____

j. big spender _____

2. Some connotations are personal and not always shared by others. If you have personal connotations for any of the following terms, write them here.

   a. lake _____

   b. ice _____

   c. summer _____

   d. chemistry _____

   e. automobile _____

PART SEVEN
ESL CONCERNS

Chapter 43 American Style in Writing

If your first language is not English, write a paragraph describing how the writing style in your native language differs from the American style in writing. Issues you might consider in your paragraph include conciseness, topic announcement, organization, and source citation. If your first language is English, interview someone whose native language is something other than English, and write a paragraph describing how that person's writing style differs from the American style.

Chapter 44 Verbs

1. The following paragraph contains some errors in the verbs. Underline the errors and write your corrections above the words you underlined.

    When people from other countries will shop in American supermarkets, they find an amazing supply of items other than grocery items. On one aisle, a shopper will discovered, for example, toiletry items such as toothbrushes, toothpaste, deodorant, and shampoo. On another aisle be paper towels, garbage bags, and cleaning supplies. Sometimes the choices would be confusing. But the choices do not ended with the shopping itself. Even when the shopping have conclude, the shopper may have decisions to make. For example, the person behind the checkout counter may asked if the customer would want the purchases in a plastic bag or a paper bag. Of course, if Americans was to shop in stores and bazaars in other countries, they can find the process just as confusing.

2. In the following paragraph underline the correct form of the verb that is needed.

    In the United States, students hope (to learn, learning) more than just the subject matter they are studying. They believe that if they (study, will study) a subject thoroughly, ask questions, and even (disagree, will disagree) with certain opinions of writers and scholars, they (will become, become) better thinkers as well. Students from other countries may find it difficult at first to practice critical-thinking skills in American classrooms because (to question, questioning) an authority or the written word may be a sign of disrespect in their country. Another problem that students from other countries may encounter in American classrooms involves writing assignments. Whereas American students are comfortable with (choosing, to choose) topics for writing that argue accepted ideas, students from other countries may have difficulty (to express, expressing) their opinions without (apologizing, to apologize) for them because (discussing, to discuss) controversial topics may be considered rude in their country. Obviously, students' cultural backgrounds affect many aspects of college life, including how they approach the educational experience.

3. Use the following verb forms in sentences of your own. If the second word can be separated from the verb in your sentence, a pronoun is included in parentheses.

   a. should + verb

   b. will be + verb

   c. may + verb

   d. can + verb

   e. do + verb

   f. may have been + verb

   g. fall behind

   h. get out of

i. look like

j. run across

k. tear (it) down

l. call (it) off

m. call (her) up

n. try (it) out

Chapter 45 Omitted Words

1. In the following paragraph there are omitted verbs and subjects. Revise the paragraph to conform to standard English by inserting verbs and subjects where they are needed.

   When importing ivory became illegal in the United States in 1989, engineers and material scientists searched for substitutes for natural ivory. For piano makers, the ban on ivory brought on the problem of how to make piano keys after the supply on hand used up. While conservationists and elephant lovers want people stop using ivory, many pianists who have tested plastic substitutes about to abandon hope of having new pianos with satisfactory substitutes. However, may be a solution. A team of experts has produced a substitute that may satisfy many piano builders and piano players. This new synthetic ivory both resembles natural ivory and has a similar microscopic structure. But even with a substitute for piano keys now available, the market for ivory not decrease as piano makers never been major consumers of ivory. The largest consumers of ivory in the world are Asian manufacturers of signature stamps, but even have given up their use of ivory and switching to substitutes.

2. Write sentences beginning with the subjects listed below. Be sure to include all verbs that are needed.

   a. My English instructor _____

   _____

   b. The other students in my class _____

   _____

   c. In my country, students _____

   _____

   d. In most schools there _____

   _____

   e. The parents of the students _____

   _____

Chapter 46 Repeated Words

In the following sentences, underline the unnecessary words that repeat the subject, pronoun, or adverb.

   a. Students in my class they are looking forward to the upcoming vacation.

   b. The book that I wanted to read it was not in the library.

   c. My teacher found the book that I left it in the classroom.

   d. The apartment building where I live there has two swimming pools.

   e. The child returned the dog that I had lost it.

   f. The pie in the refrigerator it is for tonight's party.

   g. The book drop where library books are placed there is full.

   h. My aunt called on the new cellular telephone that she had it installed in her car.

2. Complete the following sentences being careful not to repeat the subject as a pronoun before the verb and being careful not to add unnecessary words when relative pronouns and relative adverbs introduce clauses.

   a. Children in my family _____

   _____

   b. The computer that I want to buy _____

   _____

   c. The cat sat on the blanket that I left _____

   _____

   d. The car that I drive _____

   _____

   e. The girl who wants to borrow my notes _____

   _____

Chapter 47 Count and Noncount Nouns

1. The following paragraph has both count and noncount nouns. Choose the correct form in parentheses and underline it.

    In American supermarkets, new (products, product) and (produce, produces) are constantly being added to the shelves. Children are attracted to new snack foods, and adults are frequently tempted to buy items with (information, informations) about health benefits. To increase consumer (confidence, confidences) in package labeling, the Food and Drug Administration has announced new (guideline, guidelines) for various claims food manufacturers add to their labels. Products that are advertised as "low (fat, fats)" have to provide (evidence, evidences) on the label and meet new government (standard, standards). For items that appeal to children, the amount of (sugar, sugars) must be clearly indicated. Particularly helpful are the new regulations on serving size because consumers in America have become very conscious of the amount of (protein, proteins) and (fat, fats) that they eat as well as the number of (calorie, calories).

2. Write a sentence for each of the words listed below. Some are count nouns, some are noncount nouns, and some have both a count meaning and a noncount meaning.

    a. research

    _____

    _____

    b. difficulty (as a count noun) and difficulty (as a noncount noun)

    _____

    _____

    c. table

    _____

    _____

    d. computer

    _____

    _____

e. light (as a count noun) and light (as a noncount noun)

f. truth

Chapter 48 Adjectives and Adverbs

48a-b. Placement and Order

1. Reorder the following word clusters to create correct sentences, being careful about the placement of the adjectives, adverbs, and nouns. Don't add any words, and be sure to capitalize the first letter of the first word in your sentence.

   a. new   eager   three   students   entered   quickly   the   classroom

   _____

   b. Japanese   favorite   food   her   is   sushi

   _____

   c. girls   the   young   were   frightened   easily

   _____

   d. sometimes   experience   I   homesickness   intense

   _____

   e. almost   forget   never   I   anniversary   our

   _____

2. Write a paragraph of your own correctly using at least three adjectives and three adverbs. You may want to write about learning to speak a second language.

   _____

48c-d. A/An/The, Some/Any, Much/Many, Little/Few, Less/Fewer, Enough, No

1. In the paragraph below, underline the correct word in parentheses.

In northern India there is (a, an) conflict between wildlife officials and Gujjar herders of water buffalo. (A, An, The) Indian government wants to turn the area into a national park, to be called the Fajaji National Park, but for the last ten years, local water buffalo herders refuse to move off the land. The Gujjars keep herding their water buffalo, despite warnings that the animals are eating up too (much, many) of the vegetation and that soon there will be (less, few) areas that have not been destroyed by the herding. Several decades ago, the Gujjars agreed to migrate every summer to give the forests a chance to grow again, but communities in the areas they migrated to refused to accept the Gujjars because they needed (a, an, the) land for their own grazing, and they didn't have (fewer, enough, no) land to share. As a result, the Gujjars now stay in the forest throughout (a, an, the) year. Government officials keep on warning of (a, an, the) dangers of erosion in the forests where Gujjar herding has stripped the land. While (some, any, much) environmental groups say that these forest dwellers have as much right to the land as the animals, other groups support the government's attempt to move the Gujjars. The continued grazing by water buffalos risks using up the (few, less, little) food sources of elephants and other animals. Government officials plan to make (a, an) offer to the Gujjars to move them to settlements on the edge of the forests and to have them feed their animals in stalls. Park officials want to find (a, an) solution soon because they say that both the park and the Gujjars will suffer if the present situation continues.

2. Write a paragraph of your own correctly using each of the following words at least once: a, an, the, some, any, many, few, less, enough. You may want to write about how two countries vary in their concerns about wildlife.

Chapter 49 Prepositions

1. In the following paragraph, underline the correct preposition in parentheses.

    Birmingham (in, at) the spring of 1923 was a comparatively new and modern city. Lacking the Old South cultural heritage (for, of) the older southern cities such as Atlanta or Nashville, it was characterized (by, with) a tremendous energy and pursuit (on, of) success that one would expect (for, in) an emerging industrial giant. As such, it was somewhat (of, at) a melting pot (with, for) a constantly increasing number of new residents. Many were (in, from) the rural areas of the South, both black and white. There were also significant numbers of immigrants (from, in) Italy, Ireland, Germany, and the eastern European countries impoverished (by, from) war and other social and political factors. There also were smaller numbers (on, of) Asians and others. Inevitably, competition and distrust existed (with, between) these various groups, but gradually these were worked out (at, in) most cases.

2. Write sentences of your own in which you use the following prepositions.

    a. on (as a preposition of time)

    _____

    _____

    b. on (as a preposition of place)

    _____

    _____

    c. at (as a preposition of time)

    _____

    _____

    d. at (as a preposition of place)

    _____

    _____

e. in (as a preposition of time)

_____

_____

f. in (as a preposition of place)

_____

_____

g. of (to show a relationship between a part and the whole)

_____

_____

h. of (to show content)

_____

_____

i. for (to show purpose)

_____

_____

Chapter 50 Idioms

1. Write sentences of your own in which you use the following idioms.

   a. the old college try

   _____

   _____

   b. took off

   _____

   _____

   c. on the ball

   _____

   _____

   d. live high on the hog

   _____

   _____

   e. help out

   _____

   _____

   f. the bottom line

   _____

   _____

   g. come across

   _____

   _____

   h. look into (meaning to study or check on something)

   _____

   _____

i. show up

j. fall behind

45-50 ESL Concerns

Choose a paragraph of 8 to 10 sentences from a paper you have recently written or are now working on and rewrite it here, being sure that you have included all necessary subjects and verbs, that you have not repeated unnecessary words, that you have used count and noncount words correctly, that you have placed all parts of the sentence appropriately, and that you have used prepositions and idioms correctly.

PART EIGHT
RESEARCH

Chapter 51 Finding a Topic

51a. Deciding on a Purpose

For each of these topics, determine a purpose you might adopt for writing a research paper about the topic and target an audience for whom you might be writing: sports, computers, cultural differences, cloning, the media and politics.

1. Subject: college sports
   Purpose:

   _____

   Audience:

   _____

2. Subject: computers
   Purpose:

   _____

   Audience:

   _____

3. Subject: cultural differences
   Purpose:

   _____

   Audience:

   _____

4. Subject: cloning
   Purpose:

   _____

   Audience:

   _____

5. Subject: the media and politics
   Purpose:

   _____

   Audience:

   _____

51b. Understanding Why Plagiarism is Wrong

Write a paragraph explaining your understanding of
- what plagiarism is,
- what the consequences of plagiarism are in college and in the workplace,
- why plagiarism may occur,
- strategies you will employ to avoid plagiarism.

Include a description of any experience you have had or know about involving plagiarism.

51c-f. Deciding on a Topic, Narrowing the Topic, Formulating a Research Question and a Thesis

For each of the research paper general subjects listed below, narrow the subject into a manageable topic for a research paper which would be due two weeks from the date of the assignment. Then formulate a research question about your topic, and formulate a thesis statement that answers your research question.

EXAMPLE

Subject: World War II

Topic: How WWII brought more women into the workforce

Research Question: How did the economic demands of World War II affect women's roles in the American workforce?

Thesis Statement: Because World War II created an historic demand for manufacturing while simultaneously removing almost one-half of the men of working age, women entered the workforce to take up the slack, forever changing societal ideals of women's work.

1. Subject: college sports
   Topic:

   _____

   _____

   Research Question:

   _____

   _____

   Thesis Statement:

   _____

   _____

2. Subject: computers
   Topic:

   _____

   _____

Research Question:

_____
_____

Thesis Statement:

_____
_____

3. Subject: cultural differences
   Topic:

   _____
   _____

   Research Question:

   _____
   _____

   Thesis Statement:

   _____
   _____

4. Subject: cloning
   Topic:

   _____
   _____

   Research Question:

   _____
   _____

Thesis Statement:

_____

_____

5. Subject: the media and politics
   Topic:

   _____

   _____

   Research Question:

   _____

   _____

   Thesis Statement:

   _____

   _____

Chapter 52 Searching for Information
Chapter 53 Using Web Resources

52a. Choosing Primary and Secondary Sources
Identify each of the research sources listed below as a primary or secondary source for a research paper about Ernest Hemingway.

                            EXAMPLE
     "Hills Like White Elephants" (a short story by Hemingway)   <u>primary</u>

1. The Hemingway Women (a book about Hemingway)   _____

2. The Sun Also Rises (a novel by Hemingway)   _____

3. an encyclopedia article about Hemingway   _____

4. an informative television program about Hemingway   _____

5. Hemingway on the Spanish Civil War (a book of letters written by Hemingway
   and compiled by an editor)   _____

6. an Internet site about Hemingway maintained by *The New York Times*   _____

52b. Searching the Internet & 53. Using Web Resources

1. Conduct an online search for information on one of the following topics: gender equity in college sports, the relation between video games and violence, stem-cell research, or the effects of presidential campaign debates.

   Using correct MLA format, list three sources you found that you might use for a research paper on your chosen topic.

   Topic: _____

   Source 1:

   _____
   _____
   _____

   Source 2:

   _____
   _____
   _____

Source 3:

_____

_____

_____

2. Write a paragraph describing the process you used to find these sources.

_____

_____

_____

_____

_____

_____

_____

_____

_____

_____

_____

_____

_____

52c. Searching Libraries

1. Use the resources available in your college library to locate sources for a research paper on one of the other topics listed in 52b above. Using correct MLA format, list three of the sources you found. (See Chapter 60 for a complete listing of MLA formats for various types of sources.)

    Topic: _____

    Source 1:

    _____

    _____

    _____

Source 2:

_____

_____

_____

Source 3:

_____

_____

_____

2. Write a paragraph describing the process you used to find these sources.

_____

_____

_____

_____

_____

_____

_____

_____

_____

_____

_____

_____

Chapter 54 Evaluating Sources

54b. Evaluating Internet Sources

Choose one of the websites you located in the online search in the previous section of this workbook (Ch. 52b & 53). Evaluate that website using the information in Chapter 54, especially the "Checklist of questions to ask yourself about websites" on p. 348 in your handbook:

- Who is the author, organization, or sponsor? What are the credentials of this person or organization?
- What evidence is there of the accuracy of the information?
- Is the information current?
- Is there advertising on the site?
- What is the goal of the site?
- How did you access the site? Were there links from reliable sites?
- How good is the coverage of the topic?

54c-d. Evaluating Bibliographic Citations & Content

Examine the paragraph in the next exercise (Ch. 56) and its bibliographic citation. Evaluate the paragraph for use in an informative paper on world spices.

1. In this space, write your evaluation based on the bibliographic citation (author, timeliness, publisher/producer, audience).

2. In this space, write your evaluation based on the content (accuracy, comprehensiveness, credibility, fairness, objectivity, relevance, timeliness)

Chapter 56 Using Sources and Avoiding Plagiarism

56b. Summarizing without Plagiarism

Write a summary of the following paragraph. Assume that you read the paragraph as part of an article by Frederick Meisnier entitled "Historical Migrations of Foods," in the May 1993 issue of *International Cuisine,* pages 18-27.

> One of the world's most widely used spices is the chili pepper. It gives the characteristic fire to one of India's most well-known foods, curry. In Hungary, chili pepper appears as paprika, adding a crucial bite to the flavor of goulash, and in Italy the use of chili peppers in pepperoni makes the spice a common ingredient in Italian food. The hot flavor of chilies became a staple in various foods of China, Thailand, and other Oriental countries. Five hundred years ago, however, no one in these countries had heard of the chili pepper or even had a word for it, for it was one of the treasures brought back from the New World by Christopher Columbus. When Columbus returned with his new spice, its use soon spread around the globe because it was an interesting addition to the world's spice cabinets and because it traveled well in dried form. There is even evidence of chili peppers being used as a condiment in China within a few decades after it appeared on Spanish tables. The most likely route to the Orient was by sea as the Portuguese engaged in active trading with China, particularly after the founding of their trade colony in Macao. While the Spanish were the first Europeans to encounter the New World peppers, the pepper did not become a distinctive spice in Spanish cooking. It was looked upon more as a curiosity or ornamental than as a fiery way to liven up food. Now, after centuries of use in Central and South America, the chili pepper is finally becoming more popular in North American markets through the increasing interest in Indian, Indonesian, Thai, Vietnamese, and other non-European cuisines.

Write your summary here:

_____
_____
_____
_____
_____

56c. Paraphrasing without Plagiarizing

Paraphrase the first four sentences of the paragraph above.

56d-e. Using Quotation Marks and Using Signal Words and Phrases

Follow the instructions below, using this paragraph from an article by Herbert Benson, M.D., and Julie Corliss, both of Harvard Medical School, and Geoffrey Cowley, *Newsweek*'s health editor. The article is entitled, "Brain Check," and appeared in the September 27, 2004, edition of *Newsweek* on pp. 45-47. This paragraph is on pp. 46-47.

As researchers chart the health effects of hostility and hopelessness, they're also gaining unprecedented insights into the mind's power to heal. The "placebo response" has been widely recognized since the 1950s, when Harvard's Dr. Henry Beecher described the phenomenon. Until recently, most experts dismissed it as a feat of self-deception, in which people who remain sick (or never were) convince themselves they're better. But we're now discovering that expectations can directly alter a disease process. Consider those Parkinson's sufferers who improved with sham surgery. Using PET scans, researchers compared their brains with those of patients who received an active treatment. As expected, the active intervention caused a significant rise in dopamine, the neurotransmitter that people with Parkinson's lack. But the patients who improved on placebo experienced a similar dopamine surge. A related study found that fake analgesics could boost the brain's own pain-fighting mechanisms. In both cases, the placebo response was not an imaginary lessening of symptoms but an objective, measurable change in brain chemistry.

a. Quote any complete sentence from this paragraph, introducing the quote with a partial sentence of your own that would place the sentence in the context of your paragraph. Use quotation marks and an in-text citation appropriately, and punctuate the entire sentence correctly.

b. Quote a sentence from this paragraph, leaving out several words in the middle of the sentence. Introduce the sentence appropriately, include an in-text citation, and punctuate correctly.

c. Quote a sentence that includes a quote. Introduce the sentence appropriately, include an in-text citation, and punctuate correctly.

d. Paraphrase the last two sentences of the paragraph, quoting at least three words in sequence from one of those sentences. Introduce the sentence appropriately, include an in-text citation, and punctuate correctly.

PART NINE
DOCUMENTATION

Chapter 58 Documenting in MLA Style

1. Included below is a paragraph about women hikers. Following the paragraph are some sentences with information that can be added to the paragraph either as direct quotations or as information written in your own words whose source must be cited. Rewrite the paragraph so that you add at least two direct quotations and one paraphrase. Be sure to integrate the quotations and paraphrase into the paragraph, cite the sources in the paragraph, and include a list of Works Cited (in MLA format) at the end of the paragraph.

>   Among the other worlds of outdoor sports and recreation that women were entering before the turn of the century was hiking. Annie Smith Peck joined the list of great alpinists when she reached the peak of the Matterhorn in 1895. She was the first woman to wear pants for climbing, even though at that time women all wore floor-length skirts. Peck continued her climbing and hiking until she was seventy-five years old, including the historic first ascent of Mount Huascaran in Peru, a climb that went above 22,000 feet. In 1901, six years after Peck's assault on the Matterhorn, women joined the first Sierra Club hike. Despite Peck's famous knickerbocker pants, the Sierra Club's stated policy was that women wear skirts which could be no shorter than halfway from knee to ankle. Female hikers in great climbs became increasingly common, though not always as colorful as the famous Grandma Gatewood (Emma Gatewood), the first woman to hike the entire Appalachian Trail in one continuous stretch, in 1955. At the age of sixty-seven she hiked in sneakers and slept on a plastic shower curtain. Grandma Gatewood went on to hike the great Appalachian Trail two more times in her career.

Material to add to your version of the paragraph:

   a. The historian Iris Koach, in a book entitled *Women in Sports* (published in New York in 1992 by Littleton and Crane Publishing Company), wrote the following sentence which appears on page 81:

   > "Annie Smith Peck, a product of genteel Victorian wealth and exclusive private schooling, became intrigued with the idea of laying aside her scholarly efforts in order to scale the Matterhorn."

   b. On page 172, Koach mentions Emma Gatewood:

   > "With little publicity and less interest in commercial support for her efforts, Emma Gatewood continued to set hiking records until well into her eighties. She was an inspiration for women who saw themselves as too frail or inexperienced to hike on their own or without the support of a group of men."

c. The following sentence appears on page 45 of *Hiking American Trails* by William Byler (published in 1990 by Middlebauer Press in New York):

"By the 1950's, equipment for hiking filled specialty catalogs and took over more and more space in sporting goods stores. The selection of hiking boots, tents, backpacks, cooking utensils, and other gear had become a complicated science."

Your version of the original paragraph with the material added:

Your Works Cited page for your version of the paragraph you wrote in the last exercise:

_____

_____

_____

_____

_____

_____

_____

2. Listed here are a variety of sources which were used to write a paper. Using every entry listed, prepare a Works Cited page according to MLA style.

   a. A book by Martin Joos entitled *The Five Clocks* published in New York by Harcourt, Brace, and World in 1962. The paper cited information from pages 12 and 22 of this book.

   b. A book by Stephen Duggan and Betty Drury entitled *The Rescue of Science and Learning* published in New York by Macmillan in 1948. The paper cited information from pages 62-63.

   c. A book entitled *Writing Centers in Context: Twelve Case Studies,* edited by Joyce A. Kinkead and Jeanette Harris and published in Urbana, Illinois, by NCTE in 1993. The paper cited information from page 74.

   d. A poem entitled "Night Shadows" appearing in an anthology entitled *The New Gothic,* compiled and edited by Marvin Greenwood and published in New York by Chaosium in 1998. The paper cited information from page 66.

   e. A magazine article by Robert Lillo entitled "The Electronic Industry Braces for Hard Times" appearing on pages 18-23 of the February 14, 1996, edition of *Business Weekly* magazine.

   f. An interview with Nina Totenberg on the television program *Nightline,* produced by ABC and aired on channel WILI out of Chicago on November 23, 1995.

   g. A citation from William Shakespeare's *Hamlet,* taken from a collection of Shakespeare's work on CD-ROM published by CMCReSearch in 1989. The CD-ROM program does not list a place of publication.

   h. Material from an online information service, Dialog, from file 102, item 0346142. The material came from an article by Walter S. Baer entitled "Telecommunications Technology in the 1990s," published in the June 1994 edition of *Computer Science* magazine. The article began on page 152 and continued on various pages throughout the magazine.

Your Works Cited page for the sources listed in this exercise:

3. Write an informative paragraph on chili peppers by using the information in the exercise in Chapter 56b, as well as information from **two** other sources you have found through research. Incorporate information by paraphrasing, summarizing, and quoting as appropriate from all three sources and provide parenthetical citations in MLA format. List the three sources at the end of your paragraph using MLA format for references.

Works Cited

Chapter 59 Documenting in APA Style

1. Included here is the same paragraph about women hikers as used in the first exercise for Chapter 58. Following the paragraph are some sentences with information that can be added to the paragraph either as direct quotations or as information written in your own words. Rewrite the paragraph so that you add at least two direct quotations and one paraphrase. Be sure to integrate the quotations and the paraphrase into the paragraph, cite the sources in the paragraph, and include a list of the works you referenced (in APA format) at the end of the paragraph.

   Among the other worlds of outdoor sports and recreation that women were entering before the turn of the century was hiking. Annie Smith Peck joined the list of great alpinists when she reached the peak of the Matterhorn in 1895. She was the first woman to wear pants for climbing, even though at that time women all wore floor-length skirts. Peck continued her climbing and hiking until she was seventy-five years old, including the historic first ascent of Mount Huascaran in Peru, a climb that went above 22,000 feet. In 1901, six years after Peck's assault on the Matterhorn, women joined the first Sierra Club hike. Despite Peck's famous knickerbocker pants, the Sierra Club's stated policy was that women wear skirts which could be no shorter than halfway from knee to ankle. Female hikers in great climbs became increasingly common, though not always as colorful as the famous Grandma Gatewood (Emma Gatewood), the first woman to hike the entire Appalachian Trail in one continuous stretch, in 1955. At the age of sixty-seven she hiked in sneakers and slept on a plastic shower curtain. Grandma Gatewood went on to hike the great Appalachian Trail two more times in her career.

Material to add to your version of the paragraph:

   a. The historian Iris Koach, in a book entitled *Women in Sports* (published in New York in 1992 by Littleton and Crane Publishing Company), wrote the following sentence which appears on page 81:

   "Annie Smith Peck, a product of genteel Victorian wealth and exclusive private schooling, became intrigued with the idea of laying aside her scholarly efforts in order to scale the Matterhorn."

   b. On page 172, Koach mentions Emma Gatewood:

   "With little publicity and less interest in commercial support for her efforts, Emma Gatewood continued to set hiking records until well into her eighties. She was an inspiration for women who saw themselves as too frail or inexperienced to hike on their own or without the support of a group of men."

c. The following sentence appears on page 45 of *Hiking American Trails* by William Byler (published in 1990 by Middlebauer Press in New York):

"By the 1950's, equipment for hiking filled specialty catalogs and took over more and more space in sporting goods stores. The selection of hiking boots, tents, backpacks, cooking utensils, and other gear had become a complicated science."

Your version of the original paragraph with the material added:

Your References page for your version of the paragraph you wrote in the last exercise:

2. Listed here are a variety of sources which were used to write a paper. Using every entry listed, prepare a Reference page according to APA style.

   a. A book by Martin Joos entitled *The Five Clocks* published in New York by Harcourt, Brace, and World in 1962. The paper cited information from pages 12 and 22 of this book.

   b. A book by Stephen Duggan and Betty Drury entitled *The Rescue of Science and Learning* published in New York by Macmillan in 1948. The paper cited information from pages 62-63.

   c. A book entitled *Writing Centers in Context: Twelve Case Studies,* edited by Joyce A. Kinkead and Jeanette Harris and published in Urbana, Illinois, by NCTE in 1993. The paper cited information from page 74.

   d. A work entitled "Vengeance" appearing in the anthology entitled *Poetry in the Modern Age,* compiled and edited by Jason Metier and published in San Francisco by New Horizons in 1994. The paper cited information from page 54.

   e. A magazine article by Robert Lillo entitled "The Electronic Industry Braces for Hard Times" appearing on pages 18-23 of the February 14, 1996 edition *of Business Weekly* magazine.

   f. An interview with Nina Totenberg on the television program *Nightline,* produced by ABC and aired on channel WILI out of Chicago on November 23, 1995.

   g. A citation from William Shakespeare's *Hamlet,* taken from a collection of Shakespeare's work on CD-ROM published by CMCReSearch in 1989. The CD-ROM program does not list a place of publication.

h. Material from an online information service, Dialog, from file 102, item 0346142. The material came from an article by Walter S. Baer entitled "Telecommunications Technology in the 1990s," published in the June 1994 edition of *Computer Science* magazine. The article began on page 152 and continued on various pages throughout the magazine.

Your Reference page for the sources listed in this exercise:

3. Write an informative paragraph on what studies have shown about the effect of the mind-body connection on health by using the information in the paragraph in the exercise for Chapter 56d & e, as well as information from **two** other sources you have found through research. Incorporate information by paraphrasing, summarizing, and quoting as appropriate from all three sources and provide parenthetical citations in APA format. List the three sources at the end of your paragraph using APA format for references.

References

Chapter 60 Documenting in Other Styles

Choose the letter for the preferred documentation style for source citations on teach of these topics:

Topics

1. War-time journalism _____
2. Astronomical calculations _____
3. Journal of American History _____
4. Biological control of garden pests _____
5. Geometrical theories _____

Style for Citations

A. *Chicago Manual of Style*
B. *The CBE Manual* (CSE)
C. *Associated Press Style Book and Briefing on Media Law*
D. *The AMS Author Handbook*
E. *AIP Style Manual*

PART 10
DOCUMENT DESIGN & SPECIAL WRITING

Chapter 61 Document Design

61a. Placement and Order

Group the information in the paragraph below into a list that will be less crowded and have more white space.

The restaurant guide has lists of ethnic restaurants. They list Chinese restaurants in the city. They also list Italian restaurants and Mexican restaurants. As an extra feature, they list ethnic restaurants by neighborhood.

_____
_____
_____
_____
_____
_____

61b. Visual Elements

1. Put the information in the paragraph below into a **bar graph** to add interest to your Web page.

    College freshmen at the university took an average of 18 credit hours per semester. Sophomores took an average of 16 hours per semester and juniors and seniors took an average of 12 hours per semester.

2. Now put the same information in a **line graph**.

3. List all the activities that you participate in during a typical weekday, say a Tuesday, during this semester. Determine the number of hours you spend in each activity, and construct a **pie chart** that displays the information.

61c. Web Page Design

1. Draw a visual plan for a webpage for an online clothing store that would include the following pages:

   - Home Page
   - Men's clothing
   - Women's clothing
   - Children's clothing
   - Dresses
   - Toddlers
   - Children's sizes 5-10

2. If you do not already have a personal website, follow the steps below to begin planning your website.

   a. Describe the purpose for your website (e.g., entertainment, persuasion, seeking employment).

   _____

   _____

   _____

   _____

   b. Describe your home page and/or draw a replica of it here:

    c.   Draw a visual plan for your website, showing all the pages that will link to your homepage. Include as many levels of pages as you anticipate producing.

61d. Paper Preparation

Identify each of the pages below as belonging in MLA or APA style.

    1. Title Page   _____

    2. First page, with author information   _____

    3. Tables (on a separate page)   _____

    4. Footnotes   _____

    5. Works Cited List   _____

    6. Appendixes   _____

    7. End notes   _____

Chapter 62 Public Writing

62a.1 Business Letters

Write a letter recommending a person for employment. The body of your text should contain at least two paragraphs. The person to whom you are writing is Phil Brannan, New Projects Director. The address is 128 Cedar Ave., Overland Park, Kansas 66212. Be sure to send a copy to Monica Oden, Director of Human Resources. This letter should be typed or word processed and printed.

62a.2 Memos and E-Mail Communications

Write a memo or an e-mail about a situation or a problem that has arisen in the student club of which you are president. Explain the situation and propose a solution. The correspondence should be addressed to the sponsor of the club. You may write the correspondence here or type and print it.

62b. Resume Styles

1. Take the following Skills Resume and reorganize it into a Reverse Chronological Resume. The revised resume should be typed or word processed and printed out.

Donald Bryant

3256 Winslow Lane
Springfield, MO 65802
(417) 555-5300
dbryant@dotcom.net

PROFESSIONAL OBJECTIVE
    A career in hotel management that would involve guest relations, scheduling and vendor management.

EDUCATION
    Johnson County Community College
    Associate in Applied Science Degree in Hotel/Motel Management
    GPA (4.0 scale): 3.6
    Major-Related Courses:
        HOSPITALITY MANAGEMENT FUNDAMENTALS, FOOD MANAGEMENT, SUPERVISORY MANAGEMENT, HOSPITALITY LAW, SEMINAR IN HOUSEKEEPING OPERATIONS, HOTEL SALES AND MARKETING, ADVANCED HOSPITALITY MANAGEMENT, FRONT OFFICE MANAGEMENT, HOTEL ACCOUNTING

SKILLS
  Accounting
    - Student Senate Treasurer
    - Did bookkeeping for local hobby store
  Scheduling
    - Events coordinator for Student Senate
    - Scheduled work hours for College book store employees
  Vendor relations
    - Worked with various food vendors at mall food court

WORK EXPERIENCE
    Assistant Manager, Oak Park Mall Food Court; August 2001 to present Assistant Manager, College Bookstore; Fall 2000-Spring 2001

2. Type a cover letter to accompany the revised resume.

Chapter 63 Writing About Literature

63a. Ways to Write About Literature

Read a play or short story. Briefly describe each of the following aspects of the work:

1. **Plot**—How do events connect? Do some events foreshadow others?

2. **Characters**—Who is the main character in the work? Does this character change and if so, how?

3. **Narrative Structure**—Do the events proceed in chronological order? What clues does the author provide?

4. **Narrator**—Is a character telling the story? If so, how does this affect how we see the action? If not, how much are we shown (thoughts, movement in place, time, etc.)?

5. **Gender**—How does the work portray men and women? Is there a difference in how each is portrayed?

6. **Genre**—How would you describe the type of literature this work represents (give specific reasons for your choice)?

63c. Literary Terms

Define each of the following:

1. Genre _____

2. Imagery _____

3. Metaphor _____

4. Plot _____

5. Setting _____

6. Symbol _____

7. Theme _____

63d. Conventions in Writing About Literature

Next to each example below, identify the verb tense that should be used.

1. A biography of the author _____

2. A brief summary of a short story _____

3. Introducing a line from a poem _____

4. Describing the environment in which a novel was written _____

ANSWER KEY

In some cases, responses to exercises will vary, in particular when original sentences and paragraphs are to be written in response to the exercises. Other exercises require specific revision, but the method of revision may vary. In those cases, one example of a possible revision is provided in this Answer Key. Other exercises can only be revised or responded to in one specific way, and in those cases, the answers are provided here.

PART ONE

Chapter 3 Paragraphs

3a. Unity

Responses will vary but should mention that the sentence about the weather in Tibet should be deleted for the sake of unity.

3b. Coherence

Responses will vary but should mention that although the paragraph is well written, it would benefit from added transition between sentences.

3c. Development

Responses will vary but should mention that the paragraph lacks adequate development.

Chapter 14 Argument

4d.3 Logical Fallacies

    a. either…or
    b. ad hominem (or non sequitur)
    c. bandwagon
    d. circular reasoning
    e. non sequitur

PART TWO

Chapter 6 Comma Splices and Fused Sentences

Responses will vary, but one possible revision is provided here.

    Our country seems to be going through a change in attitudes. Unlike our forefathers, we Americans are now encouraged to get as much as possible as quickly as possible. This attitude is expressed on television; advertising also plays on this appeal. Many see a relationship between these current attitudes and gambling on lotteries. It is easy to understand why more and more people are handing their money over to chance, fate, and luck if what they are looking for is instant gratification. Statistics show this attitude is growing: in the late 80s, state lotteries grew an average of 17.5 percent annually—roughly as fast as the computer industry. High-tech advertising for the lotteries takes attention away from the fact that a player has virtually no chance of winning. Advertisements focus instead on the fantasy of what it would be like to win. Surveys conducted by state lotteries show that few players have a clear understanding of how dismal their odds of winning really are. A player has a better chance of being struck by lightning than of winning a lottery.

Chapter 7 Subject-Verb Agreement

1. is

    weighs

    finds

    sell

    is

    want

    depends

    is

    strike

    do

    are

are

is

2. Sentences will vary but the verb forms should be as follows:

a. Singular

b. Plural

c. Plural

d. Plural

e. Plural

f. Plural

g. Singular

h. Singular

i. Singular

j. Plural

k. Singular

l. Singular or plural depending upon how committee is used

m. Singular

n. Plural

o. Singular

p. Singular

q. Plural

r. Singular

s. Plural

t. Singular

u. Plural

v. Singular

w. Plural

Chapter 8 Sentence Fragments

1. complete sentence
2. complete sentence
3. fragment
4. complete sentence
5. fragment
6. complete sentence
7. fragment
8. complete sentence
9. fragment
10. fragment
11. fragment
12. complete sentence

Chapter 9 Dangling and Misplaced Modifiers

9a. Dangling Modifiers

The great memorials in the nation's capital are slowly deteriorating. Having faced the elements for seven decades, even casual tourists can see that the Lincoln Memorial is wearing down. Water is causing the steel structures embedded in the concrete slabs to deteriorate. At the Jefferson Memorial, which is built on clay fill, the steps are starting to pull away from the base of the monument, and there are cracks appearing in the dome. Being washed daily, details in the marble carvings of both monuments are falling off. Air pollution and acid rain are also contributing to the gradual disintegration. To find a solution, constant studies by the National Park Service are being conducted, but the caretakers of the monuments are pessimistic. Exposure to the elements, they say, causes some deterioration that cannot be stopped. Looking more like ruins than buildings, it is hard to watch these noble monuments wear away.

9b. Misplaced Modifiers

The Food and Drug Administration (FDA) claims that it has found tiny amounts of drug residue in milk samples <u>in a report issued recently</u>. The level of contamination is so low that there is no real safety problem in the nation's milk supply, but a ban has been put on one drug, sulfamethazine, found in a few samples, to prohibit its use with beef cattle and pigs <u>by farmers</u>. The FDA has already banned the use of sulfamethazine with milk cows. However, there is still concern that some dairy farmers or veterinarians use the drug improperly <u>among members of Congress</u>, even though recent testing of milk samples only found a few traces of sulfa drugs <u>in a large number</u>. The FDA is now calling on dairy farmers and veterinarians to eliminate all illegal uses of veterinary drug products. Farmers are already <u>almost all</u> complying with this order.

9c. Additional Modifier Exercises

Responses will vary, but possible revisions are provided here.

1. Scientists have discovered Cro-Magnon man was human because he used primitive tools.
2. Using a large flashlight, Bob stunned the pit bull.
3. As we walked into the house's entryway, the lamp glowed warmly.
4. Mark on the enclosed card which magazine you choose.
5. Sue found a woman's emerald bracelet.
6. As one approaches the city from the west, the sports arena is quite a sight.

Chapter 10 Parallel Constructions

Recently, two pilots, <u>one in a 175-seat commercial airliner</u> and <u>the other one in a small, twin-engine corporate jet</u>, were barreling toward each other. On the instrument panel of the commercial jet, <u>a small air traffic screen flashed a yellow circle</u> and <u>a voice announced, "Traffic."</u> As the yellow circle approached the center of the screen, it changed to a red square. The voice said loudly, "Climb, <u>climb</u>." As he noticed the red <u>square</u> and <u>as he pulled up</u>, the pilot saw the other craft fly past several hundred feet below. The voice that called out the warning was <u>not the copilot</u> but <u>the latest audiovisual aid to arrive in cockpits of planes</u>, a traffic alert and collision avoidance system. The Federal Aviation Administration has issued an order <u>that this</u>

system must be installed in 20 percent of all large commercial planes by next year and that this system must be installed in 20 percent of all large commercial planes using American airspace within the next three years. The system works by computing the distance between planes, warning planes when they get within six miles of each other, and then deciding which plane should climb and which should descend to avoid a collision.

Chapter 11 Consistency/Avoiding Shifts

Responses will vary, but one possible revision is provided here.

High school proms used to be dances where students dressed up in suits and dresses and celebrated their coming graduation, but now they have to spend a lot of money for a tuxedo or elegant formal dress. Proms have become big business as formalwear shops and limousine services offer their services in advertising campaigns. Tuxedo rental shops across the country report that proms, not weddings, account for the major proportion of their businesses. The typical expenses now include the tuxedo rental, prom tickets, corsage, dinner, and limousine rental. High-school graduates can easily spend $300 on the dance, and that may be just for the basics. Prom-goers can spend additional funds for the latest fashions in tuxes, and they can buy photos for $50 or more. Even when the prom is over, there are other expenses. Some graduates go away for the whole weekend, often to a resort hotel. So there are additional costs for a hotel, and if parents are along as chaperones, it is necessary to add in the cost of their rooms and meals, too. High-school graduates insist that this is all necessary as a rite of passage, but parents who often have to contribute at least part of the funds do not appreciate the expense.

Chapter 12 Faulty Predication

With the present concern for the environment, some companies are trying to increase their sales by advertising their products as environmentally safe. The makers of some plastic trash bags, for example, are claiming that their plastic is degradable. The reason for the claim is that there are additives that cause the product to break down after prolonged exposure to sunlight. Biodegradability, they assert, occurs when there is photodegradability, a breakdown by sunlight. But since most trash bags are buried in landfills, the benefits of photodegradability are questionable. Thus, the government has stated that one way to improve deceptive advertising claims is to eliminate false or misleading information.

Chapter 13 Coordination and Subordination

Responses will vary, but one possible revision is provided here.

Americans have become videotape addicts. Because over 75 percent of all American households now have videocassette recorders, hotels hope to make their guests feel at home by providing equipment for watching videotapes. Some of the largest hotel chains in the country have added videocassette recorders in the rooms which can be used by renting videocassettes from shops in the hotel. In these shops, which have a good selection of recent movies for guests to view in their rooms, hotels guests can show their room key and can rent a movie at a reasonable rate. Other hotels don't want to bother with stores in the lobby, and they are exploring a different option. They are adding automated video dispensing machines in their lobbies that hold hundreds of titles and have new releases as well as standard favorites. One hotel chain, which has a lot of large business conventions, has investigated another approach. It is offering to distribute to guests videotapes that the corporation holding the convention wants its participants to see. Corporations like this because the videotapes can convey some of the key ideas being presented at the convention. Videotape players in hotel rooms may soon be standard equipment, just as television sets were added years ago. When Americans became addicted to television, they expected to watch television in their hotel rooms.

Chapter 14 Sentence Clarity

Responses will vary, but one possible revision is provided here.

A new approach to using computers may be to jot handwritten notes onto an electronic pad. This is done by a special pen that projects a narrow light beam onto the pad. For computer users who have had to rely on entering data into a computer by means of a keyboard, it is a major step forward in using computers. Typing is more distracting than writing for many people who are not skilled typists but who use computers frequently. For the last twenty-five years, the goal of computer developers has been to rid the computer of keyboards, but more development in the field of character recognition is needed. One way to eliminate the keyboard is to teach computers to read through optical character recognition. Another way is to recognize the human voice instead of typed data. But speech recognition is not advancing as rapidly as some computer developers would like. It is not likely that the near future will see computers we can converse with because we have no appropriate technology that is this advanced. More promising is the ability of the computer to scan images. Already character recognition machines scan pages of printed material. Computer developers are also considering electronic gloves that could be used by people to point to areas of the screen. There is no limit to what will be coming next in computer development. The result of doing away with the keyboard will be to save time and to eliminate all those typos.

Chapter 15 Transitions

Responses will vary, but one possible revision is provided here.

My father really needs help because he is a workaholic. He works nearly twelve hours a day. Doctors have told him it is not only bad for his health, but it is affecting his family, and he has worried my family to no end. My family has tried to help him, but he has not acknowledged the problem. He has a tolerance that goes through the roof. He sneaks out of the house before anyone is awake, and he gets home after most of us have finished dinner and are getting ready for bed. He works at home as well: his computer is connected to the one at his office, and his work calls him at home. He has a pager and a cell phone. Things may be getting worse since Dad accepted a promotion to regional director.

Chapter 16 Sentence Variety

1. Responses will vary, but one possible revision is provided here.

   Animal rights activists, who are known primarily for their campaigns against fur coats and the use of animals in laboratories, are also campaigning against rodeos. In their public statements seeking support, protesters say that rodeo animals are being mistreated. For example, they say the animals are starving and live a life filled with pain and suffering. Although animal rights activists also say that rodeo horses buck because they are in pain, rodeo and circus owners say that this is not so. Well-fed and comfortable, these animals would be going to slaughter if they were not used as show animals. Moreover, to emphasize their point, handlers explain that they have a healthy respect for the size and power of these animals. Treated like star athletes, the animals enjoy performing. On the other hand, animal rights activists point to the use of cattle prods and bucking straps to get the animals moving, but rodeo owners argue that even ranchers have to use electric prods to get herds moving. Before prods were used, ranchers used pitchforks. Calf roping, also condemned by animal rights activists, can break calves' necks and can also snap vertebrae and legs. Calf roping, to the satisfaction of animal rights activists, has been banned in some states. Eliminating calf roping, they hope, can result in the overall elimination of rodeo shows. However, rodeo owners protest that breaking of calves' necks, bones, and vertebrae just doesn't happen, and they invite activists to come to rodeos and see for themselves.

3. Responses will vary, but one possible revision is provided here.

Gardening can be a worthwhile activity because the gardener can produce food for her family, as well as helping to save the environment and prevent the overuse of landfills. One way to do this is by mulching, which is actually recycling. The gardener can recycle clippings when she mows the lawn in the summer and leaves when she rakes in the fall, along with scraps from the kitchen all year long. Usually the clippings, leaves, and scraps are gathered in large plastic bags and taken to the landfills. To use these items as organic mulch, the gardener should collect them in a corner of the garden and mix them all together. Later, she can spread the mulch on the garden to provide nutrients which will cause the soil to produce bigger and better vegetables, thus providing her with many benefits.

PART THREE

Chapter 17 Verbs

17a. Verb Phrases

   Some scientists <u>are saying</u> that a buildup of carbon dioxide and other greenhouse gases in the atmosphere <u>will cause</u> global warming. But another group of scientists <u>argue</u> that we <u>should study</u> the data more carefully before any firm conclusions <u>are drawn</u>. While scientists generally <u>agree</u> that an unchecked accumulation of greenhouse gases <u>will cause</u> changes, no one <u>knows</u> when it <u>will start</u>, how much <u>will happen</u>, or how rapidly it <u>will occur</u>. The most widely accepted estimate is that there <u>will be</u> a rise in the earth's average temperature as early as 2050. This <u>could bring</u> rising sea levels and severe droughts in some areas. But no one <u>knows</u> yet how clouds and the ocean's ability to absorb heat <u>will affect</u> this. When scientists <u>understand</u> this better, projections <u>can be revised</u>.

17b. Verb Forms

   The evidence that global <u>warming</u> has <u>started</u> is not very strong. Some scientists believe that the concentration of carbon dioxide has <u>increased</u> over 25 percent since the early 1800s, but other scientists point to the fact that the average global temperature has <u>risen</u> by no more than a half degree Centigrade. Even that rise is questionable since there was a <u>cooling</u> period from 1940 to 1970 that caused forecasters <u>to predict</u> a return to the ice ages. Therefore, <u>to act</u> on predictions by <u>passing</u> laws that restrict or ban the use of fossil fuels may be hasty, but <u>conserving</u> energy, <u>banning</u> harmful chlorofluorocarbons, and <u>planting</u> more trees <u>to absorb</u> carbon dioxide from the air seems sensible. Many industries are also <u>acting</u> more responsibly and are <u>reducing</u> hazardous emissions from their factories.

17c. Verb Tense

1. In 1987, the FBI (<u>opened</u>, has opened, was opening) a $1.5 million mock-up of a small town in Virginia named Hogan's Alley. Presently, the town's population (has been, will be, <u>is</u>) about 200. Generally, Hogan's Alley (has looked, <u>looks</u>, is looking) like any peaceful little American town, but often the quiet (will be shattered, <u>is shattered</u>, had been shattered) by the sound of a shotgun or squealing tires. Then prospective G-men who (<u>had been sitting</u>, sat, will sit) outside the post office jump in their cars and race after the "criminals." None of this (will be, <u>is</u>, is being) real because Hogan's Alley (<u>is</u>, has been, had been) a training academy for those who (will want, <u>want</u>, will have wanted) to become FBI agents. Next year over 500 trainees (will have attended, attend, <u>will attend</u>) classes in frisking, lectures on handcuffing, and seminars on interrogating witnesses. The FBI (<u>started</u>, was starting, had started) Hogan's Alley because it wanted its agents to have more true-to-life experience before they go out and deal with dangerous criminals. So now every day there (will be, have been, <u>are</u>) mock bank robberies, kidnappings, and drug busts. The "criminals," however, are actors, part-time students, retirees, off-duty policemen and firemen, and anyone else who (<u>has passed</u>, will pass, had passed) the rigorous screening tests.

2. Last year, in Hogan's Alley, the FBI's mock-up village for prospective FBI agents, the trainees <u>took</u> part in an intensive 14-week training course. About 46 agents <u>moved</u> through the program which <u>began</u> with lessons in surveillance. Trainees <u>tracked</u> a suspect from his home as he <u>drove</u> to a shopping mall and <u>sold</u> a small bag of phony cocaine. Next <u>came</u> a class in simple arrests, when agents-in-training <u>burst</u> into a fleabag motel and <u>took</u> an unarmed burglar as he <u>lay</u> in bed. Agents <u>learned</u> how to frisk suspects, and then they <u>read</u> them their rights. Instructors, who <u>brought</u> their own experience as former FBI agents to the teaching, carefully <u>gave</u> advice to each trainee and <u>provided</u> pointers on how to handle details such as slipping on handcuffs during a struggle. Nine weeks into the course, trainees <u>got</u> some practice in arresting an armed felon. Their final lessons <u>took</u> place in a courtroom, where they <u>faced</u> a team of actors who <u>posed</u> as a tough team of defense lawyers. Those trainees who <u>did</u> not flunk out of the course <u>became</u> FBI agents, and the

actors, who played drug peddlers, burglars, and other felons, said that seeing what it is like being on the wrong end of the law reminded them that they did not want to become criminals themselves.

17d. Verb Voice

Like many cities built on a river, Kansas City is divided (passive) between two states. The biggest single municipality is Kansas City, Missouri, but over one-half of the population of the metropolitan area lives (active) in various suburban cities on the Kansas side. Problems are created (passive) when it comes time to finance projects that will benefit (active) both sides of the state line. In the 1990s, cities on both sides of the state line agreed (active) to a bi-state sales tax to be used to renovate Kansas City, Missouri's Union Station. Advocates said (active) the new station would bring (active) in tourism that would benefit both sides of the state line. Opponents claimed (active) Kansas City, Missouri was just trying to dip (active) into the pockets of the more affluent suburbs on the Kansas side. In the end, a compromise was reached (passive). The bi-state sales tax was passed (passive), but all contributing cities had (active) a say in its use and the tax had (active) a 5-year duration. It was not renewed (passive).

17e. Verb Mood

Singles bars and dating services are thriving (declarative), but there are always new approaches. In commercial dating services, one new approach that may be cheaper (subjunctive) than the standard videotaped interviews is the lunch-date service. For less than $50 a month, the company promises (declarative) three lunch dates a month. People are paired on the basis of simple criteria gathered from brief interviews that might last (subjunctive) less than five minutes. The company sets up the lunch date, and the participants take it from there. "Meet (imperative) new people," says the advertising brochure, "and enjoy (imperative) some interesting little restaurants." Since the Census Bureau puts (declarative) the number of single adults at 66 million and growing, these new twists on dating services may prosper (subjunctive).

17f. Modal Verbs

When American products began appearing in Moscow, city officials worried that the signs and advertising for these products <u>might</u> make (<u>could possibly make</u>) Moscow look less Russian. A law recently passed in Moscow warns that all stores and businesses <u>must</u> display (<u>need to display</u>) signs in Russian or at least change them into the Cyrillic alphabet. This has caused many businesses to contact the city inspector because of questions they have. For example, one businessman wondered whether he <u>should</u> change (<u>is obliged to change</u>) the letters in the label for the Puma running shoes that he sells. He was concerned that changing the letters from English to Russian <u>might</u> make (<u>has the possibility of making</u>) the shoes less well-known. The problem became confusing because the new law does not forbid foreign words but does require the Russian sign to be bigger than the one in English. Different stores found different solutions. An American cosmetics company, Estee Lauder, announced that they <u>will</u> put (<u>strongly intend to put</u>) one sign with the Russian alphabet on their awning and another sign with the English alphabet in the windows. Despite all the questions and worries, the inspector in charge of enforcing this rule will be very strict about checking on foreign signs.

Chapter 18 Nouns and Pronouns

18a. Nouns

Sports fanatics are spending large sums of money these <u>days</u> to purchase sports memorabilia. In fact, says one of the <u>industry's spokespersons</u>, it is a $100 million-a-year obsession. Buyers look for scorecards, stadium seats, autographed baseball bats, and even <u>children's</u> clothing that belonged to old-time greats such as Babe Ruth. <u>Ruth's</u> personal letter to a fan recently sold for over $7000. Since Joe Lewis, the famous boxer, kept many of his old <u>mouthpieces, collectors</u> are paying over $2000 for boxed sets of Lewis's mouthpieces. There is even a magazine entitled *Sports Collectors Digest* which keeps <u>collectors</u> informed about many of the <u>auctions</u> and the availability of various <u>items</u> of interest. A faculty member in the history department of a major American university has begun a study of this phenomenon and reports that yearly increases in prices are staggering. For example, he says, a nineteenth-century baseball card of an obscure Hall of Fame pitcher,

Tim O'Keefe, which was worth about $750 two years ago, recently sold for over $30,000. O'Keefe's other memorabilia are expected to soar in value as well. Sports collectibles are now big business.

18b. Pronouns

The business of sports collectibles has become so profitable that it has attracted con artists who manage to forge and sell bogus items. The forgeries have become such big business, in fact, that many con artists help one another and have developed a large network of bogus items. These items include fake Joe DiMaggio signatures and imitation press box pins. Anyone who is in the memorabilia business can spot these forgeries and fakes, but sports fans are often too enthusiastic about a purchase to take the time to have their purchases checked by experts. As a result, they often allow themselves to be conned. A lot of junk from people's attics is also cluttering the market. When someone finds yellowed pages from 1929 sports sections in her scrapbook, she may think she has an expensive treasure. Flea markets overflow with these antiques which may or may not be worth anything. But if a buyer is willing to pay large sums of money, who can say whether that autographed photo or threadbare jersey is or is not a treasure worth collecting? With a market where buyers pay $20,000 for one of Lou Gehrig's old bats, it is worth cleaning out those old attics and scrapbooks.

Chapter 19 Pronoun Case and Reference

19a. Pronoun Case

1. When historian Page Smith began to ask himself just what kind of education was being provided by American universities, he started with a list of questions. His answers can be found in his book entitled *Killing the Spirit.* Those readers looking for encouraging conclusions will not find them in this report as Smith criticizes professors' lack of interest in teaching and their tendency to spend too much time writing books and articles that do not contribute anything useful to mankind's betterment. Smith finds those academics who look down on teaching as less important than research to be at fault. For him, the student revolt of the 1960s and early 1970s was a time when students were looking for answers to important

questions but got no help from their professors. There was no dialogue between their professors and them that addressed questions about meaning and values in life, says Smith. However, community colleges and small, independent colleges still concern themselves with the needs of students. University administrators who are more interested in budgets and specific agendas will not find answers in Smith's book, and readers will wonder who his real audience is. To whom should we look for answers?

3. Fall break was over, and the research assignment was due the next day, so Ross asked Lucy if she wanted to go to the library with Amy and him. The instructor had told the class that the students who made high grades on the midterm exam would be those who had read the assignments and completed all the homework. When the girls and he reached the reference room, they saw several classmates whom they could work with. They asked their friends, "Who has finished the assignment?" They were surprised that all the other students were finished and headed to Starbucks. It was clear that a long night lay ahead for the girls and him. The next day, Ross asked the teacher, "Please give Lucy, Amy, and me one more day to finish the research assignment." But the teacher said that wouldn't be fair to the rest of the students who had all worked hard, so Ross and they learned, yet again, not to procrastinate.

19b. Pronoun Reference

When Benjamin Franklin discovered electricity in thunderclouds, he sparked a controversy that still has no clear answer. How do clouds become electrified in the first place? To this day they haven't been able to adequately explain how it contains such incredible amounts of electricity that a stroke of lightning can contain about 100 million volts. One researcher who wants to find some answers flies his plane into storms to measure electric fields and ice particle changes. It's bumpy work, he says, especially if there are large hailstones because it can damage the plane and the measuring equipment. On one trip they noticed that in sections of clouds where water and ice mix, the measuring devices picked up indications of strong charge separation. The answer may be that in a certain temperature range, the temperature can cause the charge separation. Another factor may be a kind of soft hail called "graupel," pea-sized particles that can look like miniature raspberries. They form when droplets of supercooled water collide, freezing together instantly. Ice crystals then bounce off the growing graupel, building up a charge from the friction just as you build up a static electricity charge when you scuff your feet across the carpet. When they are carried to different parts of the cloud, the result is a separation of the positively and negatively charged particles. Then, when the electrical difference between the ground and the sky becomes great enough, everyone should haul his or her kite in.

Chapter 20 Adjective and Adverbs

20a. Adjectives and Adverbs

For many people, the crossword puzzle in the daily paper is one of life's little pleasures. Some say it is also surely one of life's frustrations. While puzzle books conveniently include the answers in the back, most newspapers print the answers the next day. Now there is a quicker answer. Some companies have an automated solution. Readers can dial an 800 service for instant answers. This service is free to callers but is paid for by advertisers who sponsor each day's puzzle. An advertiser can run a small advertisement by the puzzle and can include a ten-second message that callers will hear before the answers are given. The New York Times handles this differently. For their puzzles, callers pay for all requests for clues. These services are a major breakthrough for frustrated puzzle-doers who are used to waiting until the next day.

20b. A, An, The

After winding down twenty miles of dirt road in Mtunthama, Malawi, (no article) visitors will come upon the well-kept lawns and gardens of (the) Kamuzu Academy, one of (no article) Africa's most unusual schools. Here, on four hundred acres of well-trimmed lands is (a) school dedicated to classical scholarship. (The) school was founded by President Hastings Kamuzu Banda, the ruler of Malawi since it won its independence from (the) British in 1964. Originally, Dr. Banda studied at (a) Scottish missionary school in Malawi and went on to study in (no article) South Africa, (the) United States of America, and (no article) Britain, where he acquired (a) medical degree and (a) love of Latin and Greek, as well as a strong attraction to the classical emphasis of the elite British schools. When Dr. Banda returned to Malawi, he wanted to copy (the) architecture and curriculum of Eton and other British boarding schools. Some critics in opposition to Dr. Banda say that (the) academy is not appropriate in their country where ordinary schools often do not have (no article) textbooks and where more than two-thirds of (the) population are illiterate. However, defenders of the school point out that the school is not elite in choosing its students. Children are accepted without regard to their family's wealth or position. Each year 35,000 students take (an) exam to try to gain entrance to the school which accepts about eighty new students (a) year. Once accepted, all students are required to take four years of (no article) Latin and four years of (no article) ancient Greek, along with (no article) English, (no article) mathematics, and (a) history course about Africa. Most graduates go on to university and then take jobs in (the) Malawi civil service.

20c. Comparisons

1. larger
2. largest
3. more exciting
4. most enjoyable
5. user-friendliest (or most user-friendly)
6. better
7. best
8. worst

Chapter 21 Prepositions

Americans used to pour ketchup over so many foods that it was America's favorite condiment. Now salsa has replaced ketchup as the favorite between the two in American food markets. Salsa, which means "sauce" in Spanish, is defined as any fresh-tasting, chunky mixture, usually made with tomatoes, chilies, onions, and other seasonings. Although salsa used to be associated with Mexican and southwestern cooking, it is being used for a variety of foods not particularly Mexican. The popularity of salsa is apparently part of the current food trend as Americans become more interested in spicier foods. In 1988, only 16 percent of households in America bought salsa. In two years, that figure was up to 36 percent, and the market continues to grow at a very fast rate. A marketing information company notes that salsas and picante—which are different from salsas because they are thinner—account for about two-thirds of this market, a category that also includes taco and enchilada sauces. As the market expands, so do the choices. The simplest salsas are based on chopped tomatoes and chilies, and the types of chilies determine how hot the particular type of salsa is. Cookbooks with a variety of recipes indicate the preference of some people to make their salsas at home. Salsas are one of the few popular snack foods that are fat-free or nearly so, and many are made without preservatives, two characteristics that may contribute to their popularity now that people are interested in eating healthier. Riding on this wave of popularity are the new fruit salsas, made with peaches, pineapples, and so on, and vegetable salsas, made with pinto beans, corn, and black-eyed peas. Other varieties will surely appear as the market expands in the future.

Chapter 22 Subjects

Because television viewers of sporting events like to follow the actions of specific players, a cable company in New York is trying out a new subscription service. If it is successful, it is likely to become available to everyone. Cable subscribers will be able to tune into a tournament and pick the player they want to follow. Using a wand similar to a TV remote control, viewers sitting at home will be able to switch from conventional network coverage to a camera that focuses solely on the preferred player. Instant replays and a list of statistics on selected pros will be other selections that the service will provide. This interactive technology will allow viewers to customize programming. While the present programming is only a test of the technology, some cable companies foresee customers being able to sign up for interactive television just like other cable services. When will this service be available for everyone? It is too soon to know for sure, but there is certainly a large demand for such services.

Chapter 23 Phrases

1. c
2. b

3. f

4. c

5. e

6. a

7. d

Chapter 24 Clauses

24a. Independent Clauses

If you are a magician and like highly structured organizations, <u>you may want to join either the International Brotherhood of Magicians or the Society of American Magicians</u>. If, however, you are a magician who lives in New York City and likes to gather informally with other magicians, <u>you should visit Reuben's Restaurant on 38th Street and Madison Avenue any Saturday afternoon</u>. <u>There will be professional and amateur magicians sitting in the back room</u>, and <u>they will be swapping tricks or polishing their routines on each other</u>. <u>People come in with jumbo interlocking rings, decks of cards, and brightly colored scarves in their pockets</u>. <u>Some are doctors, professors, salespersons in shoe stores, tax attorneys, shipping clerks, or teenagers</u> who are just learning the basics. <u>Others are professionals</u> who have appeared on *The Tonight Show* or in traveling circuses. <u>They all share a love of magic</u>, and <u>they willingly sit and watch each other's routines</u> because they know how valuable it is to keep practicing.

24b. Dependent Clauses

There are an estimated 50,000 magicians in America. Most are amateurs <u>who enjoy magic as a hobby</u>. These amateurs often have elaborate equipment <u>although their only audience is usually their friends and relatives</u>. Some, however, specialize in the small card, coin, and rope tricks <u>that are always popular</u>. Purists call this intimate "close-up" magic the only real magic <u>because it relies so heavily on a person's manual dexterity</u>. <u>Because people seem to prefer to be fooled face-to-face</u>, this close-up magic is also offered by professional magicians who perform at birthday parties and trade shows. A psychologist who is also a magician says that when something is done under people's noses, it's more magical. It's much more elusive. The spectacular effects of magic done on television don't seem to impress people quite so much. Whatever the cause may be, amateur magicians will keep buying those sponge balls, decks of cards, special coins, and paper flowers.

Chapter 25 Essential and Nonessential Clauses and Phrases

Responses will vary, but one possible revision is provided here.

    Music teachers in the elementary schools who want to teach classical music to their students now have a lot of materials to help them. The materials have been developed by a group of educators who want to introduce classical music appreciation programs in elementary schools. Many students say these materials are excellent. These programs are also being used to coach students for contests. The programs incorporate listening skills and information about themes, styles, and forms of music. Students seem to enjoy these contests. The contests are a new form of competition in many schools. Contestants are often asked to identify the main themes of familiar works such as *Nutcracker Suite,* the specific instruments being played, and the composers. Harder questions are those that are asked about music the students have not studied in school. These questions stump even the more experienced teams. Not all the music that is studied in the music appreciation classes is classical. Sometimes jazz and non-Western music is added, though students also seem to enjoy studying classical music that they have heard on television commercials. Music teachers who have tried these programs are pleased with the variety of types of music students come to like and the quantity of music they come to know. These music appreciation programs are expanding rapidly across the nation. The programs were begun on a small scale in a few states.

Chapter 26 Sentences

1. c
2. c
3. e
4. b
5. e
6. c
7. a
8. a
9. e
10. d
11. c

PART FOUR

Chapter 27 Commas

27a. Commas in Compound Sentences

In 1977 the flight of the little airplane, the Gossamer Condor, did not look very impressive, but it was indeed an historic flight. With wings of foam, balsa wood, and Mylar, the plane designed by Paul MacCready floated slowly and gracefully over the San Joaquin Valley and covered a mile or so in about eight minutes. What made it so historic was that the pilot was pedaling. The Gossamer Condor was only the first of MacCready's pedal-powered planes, and two years later the plane's successor, the Gossamer Albatross, crossed the English Channel. Some people say that MacCready is really the brains behind these inventions, but others feel that he receives undue credit for the work that others on the development teams do. However, MacCready was the first to use his observations of how birds fly, so his supporters feel that he is the genius who made human-powered flight possible. Other inventors were taking the conventional approach of trying to reduce drag as much as possible because they thought this approach would be the answer. The approaches of other inventors were to streamline their aircraft, or they tried to incorporate ways to increase the horsepower. Only MacCready applied the principle of vastly increasing the wing area and used materials to keep the overall weight down. The result was an aircraft that needed only the power output of a good bicyclist, and the Gossamer Condor now has a place of honor next to the *Spirit of St. Louis* in the Smithsonian Institute's National Air and Space Museum.

27b. Commas After Introductory Words, Phrases, and Clauses

Having won prizes with his first human-powered plane, Paul MacCready went on to build a faster and more powerful pedal-powered plane, the Gossamer Albatross. In less than two years after his first success, MacCready's second pedal-powered plane departed from Folkestone, England, in June, 1979, bound for France. Expecting the flight to take about two hours, MacCready allotted just enough water for the pilot to drink. The flight team who prepared the plane and assisted the pilot on the ground waited several weeks for the kind of calm weather that was needed. Consequently, the pilot took off, expecting to reach France before his endurance and the water gave out. But a head wind blew up soon after the pilot was aloft. An hour and a half later, he was only two-thirds of the way to France, and his legs were cramping from all the pedaling. Because everyone was sure they had to give up the attempt, the flight team was ready to hook a towline to the craft that would haul it ashore. Tired and about to give up, the pilot knew he had to gain altitude to get hooked to the towline. As he climbed, he found less wind and was able to press on. Almost three hours later, the pilot touched down at Cape Gris-Nez, in France, a minute short of his theoretical exhaustion point. The Gossamer Albatross had crossed the English Channel, powered only by the pilot.

27c. Commas with Essential and Nonessential Words, Phrases, and Clauses

After designing human-powered planes, Paul MacCready, a prize-winning inventor, went on to design a solar-powered plane. MacCready, however, realized that solar cells as an energy source for planes do not make any practical sense. But MacCready, who had long sympathized with environmental concerns, hoped to demonstrate that solar power has an important part in the world's energy future. Those who see solar energy as merely a minor source of energy for the future downplay the importance of such demonstrations. Others think solar power has simply not been adequately developed for practical use. The solar-powered plane that MacCready designed flew from Paris to the coast of England in 1981, cruising at 441 mph at an altitude of 11,000 feet. The plane, called the Solar Challenger, provided the stepping stone to MacCready's next flying machine, the Sunraycer, a solar-powered car.

27d-h. Commas in Series and Lists; Commas with Adjectives; Commas with Dates, Addresses, Geographical Names, and Numbers; Other Uses for Commas; Unnecessary Commas

The Sunraycer, which is a solar-powered lightweight car, was built to compete in the 1987 Race Across Australia. Designed by Paul MacCready, the car won the race from Darwin to Adelaide_ and is now in the Smithsonian Institute's National Museum of American History. With a total weight of 365 pounds, the car has a power output of about 1.8 horsepower at noon on a bright day, and it gets the electric power equivalent of 500 miles to the gallon. The Sunraycer, which presently holds the solar-powered speed record of 48.7 mph, averaged a little over 40 mph for much of the race. The car is so light_that when it made turns during testing, it often seemed in danger of blowing over. The engineers who worked on the Sunraycer_ended up putting two little ears on the top. "We're not sure why they work," said one engineer, "though they seem to help." In some ways the Sunraycer is not a prototype of electric cars for commercial use_ because the Sunraycer has bicycle-thin wheels, a driver's seat that requires the driver to lie flat, and very weak acceleration. However, many of Sunraycer's features were carried over into the electric car developed by General Motors. Like the Sunraycer, the GM car uses alternating current_ and can, therefore,_get better performance. If electric-powered cars become widely popular in the future, Paul MacCready will feel that he had some small part in saving the environment.

27a-h Review of Comma Usage

1. address
2. no commas
3. appositive
4. items in series
5. nonessential clause
6. introductory element
7. introductory element, date
8. no commas
9. compound sentence
10. interrupter, contrasting element

Chapter 28 Apostrophes

In the game of baseball, batting slumps are one of a player's worst nightmares. When they are doing well, players attribute their successes to mysterious minor occurrences around them that then become habits_ the players keep up. After a game in which one baseball player who was wearing an old helmet with its side dented hit two triples, a home run, and a single, the player continued to wear that helmet for the rest of the season. Warding off evil spirits through superstitions is another thing baseball players do. One player always wears the green T-shirt from his university under his team's jersey. Another won't wear a jersey with any 6's in his player's number. Batting coaches spend hours watching videotapes with slumping players, trying to find what's causing the problem. They examine the player's batting stance or swing, but this doesn't always provide useful clues. Some slumps happen when batters begin to worry too much about their misses and about everyone else's successes. But one coach thinks otherwise. He notes that some players start making adjustments when they've hit a double and want to hit farther or when a certain unusual pitch connected well with their bats. Players in their 30's complain of a different kind of slump. One bad day, says one over-30 player, may mean he is losing it, that his age has begun to take its toll. A hitter who is hot, on the other hand, tends to get possessive about a special bat and uses it until it breaks or its cracks begin to show. When that bat goes, the player sometimes loses confidence until he connects with a new bat that brings him a few homeruns. A batter's life isn't as easy as some people think it is.

Chapter 29 Semicolons

Junk mail used to be confined to print on paper; now it is appearing in people's mailboxes on videocassettes. Companies in the direct-mail business are now marketing inexpensive cardboard videocassettes that can carry a variety of messages, such as audiovisual advertisements, promotional premiums, and educational or training aids. Some companies are switching to this form of direct-mail advertising because it is relatively cheap; in addition, it presents messages more vividly on television screens than print advertising can on paper. Informational videos can be sent to prospective customers, and advertisers can use their product in the video. Says the spokesperson for one cereal company, "We are interested in promoting good nutritional habits"; as might be expected, the balanced diet they will be picturing in the video will include their cereal. Printed instructions in merchandise are often confusing; consequently, some manufacturers are also switching to these disposable videocassettes for the instructional packets they include with their merchandise. The cardboard videocassettes are relatively cheap to manufacture and cheap to mail. They are certainly more convenient than the promotional packets sent by companies that have relied on enclosing sample packets of toothpaste, aspirin, or cereal; mail advertisers who send bulky envelopes of coupons; and companies who want to entice customers with big brochures of vacation places, hotels, and tours. The disposable videocassette is certainly going to grow in popularity as an advertising medium.

Chapter 30 Colons

Evaluating college teachers is a complex, often difficult task. Now a new method devised in the Midwest is being tried out across the nation: the teaching portfolio. Administrators and faculty members see the portfolio, a collection of materials documenting classroom performance, as a way to emphasize teaching as a major priority. Too often, professors who are being evaluated document primarily their research and scholarly activities, such as published books and articles, lists of grants, and conference presentations. Says one faculty member at a large Eastern university: "When it comes to teaching, most teachers only have student evaluations and coffee-room conversations about what they do in class and how their students are learning"; this kind of evidence is not very thorough or sufficient to determine pay raises, tenure, and promotion. As a result, many faculty members are judged and rewarded on the basis of their research and publications. Their teaching performance, whether outstanding or mediocre, is largely ignored. The portfolio can be a better measure of what faculty really do as teachers, for the portfolio can consist of the following items: a statement of the person's teaching philosophy, a list of the courses that person has taught, a representative syllabus, statements by fellow faculty members who have observed this person, and course materials that the person has prepared. At some schools where portfolios are already in use, the portfolios also include unsolicited letters from former students, teaching awards, and even videotapes of the candidate teaching a class. Some faculty members who have prepared these portfolios say that there is an added advantage: the portfolio collection can lead to self-improvement. This can happen when the process of creating the collection causes the person to think seriously about teaching goals, strategies, and results.

Chapter 31 Quotation Marks

"Why am I fatter than my sister-in-law? I eat less," complained a woman being studied by a team of researchers. She explained that she repeatedly went on diets when her sister-in-law didn't. But the woman continues to weigh more. Researchers are finding out that heredity, in addition to lifestyle, exerts a strong influence on people's weight. By studying identical and fraternal twins, research teams are finding that brothers and sisters end up with similar body weights whether or not they are raised in different families. In the *Journal of Genetics*, Dr. Albert Skinnerd writes, "When the biological parents are fat, there is an 80 percent chance that their children will also be overweight" (234). "Does this mean that my brother and I are doomed to be fat?" asked one overweight twin in the study. Since some sets of twins tend to transform extra calories into fat while other sets of twins tend to convert extra calories into muscle, one scientist concluded that "genes do seem to have something to do with the amount you gain when you overeat." Some unsuccessful dieters may be relieved to know that their failed diets aren't a matter of failed willpower. "It is really a matter of metabolism," reports another doctor doing research in this field. But that does not mean that low-fat diets and exercise should be given up. " 'Quit' is not a word in my vocabulary," says one constant dieter who manages to maintain a reasonable weight by means of careful eating and plenty of exercise, despite a tendency to be overweight.

Chapters 32, 33, 34 Hyphens, End Punctuation, and Other Punctuation

1. Which oils are good for us to eat? A study of thirty-nine participants on a reduced-fat diet looked at the benefits of consuming olive oil and/or corn oil. Which is more beneficial in influencing high-density lipoprotein levels— corn oil or olive oil? This is an important question because high-density lipoprotein (HDL) is considered a beneficial form of cholesterol that helps remove the more dangerous low-density lipoproteins from the body. In a study published in an article, "Two Healthy Oils for Human Consumption" (*Diet and Health News* 14 [1991]: 22-36), researchers report that the participants first spent twelve weeks on a diet that included olive oil and then another twelve weeks on a corn-oil diet. The results (which were also announced on television newscasts) indicated that neither diet resulted in lower levels of HDL. This indicates that a diet of olive oil and/or corn oil can accompany a

reduced-fat diet. For those who self-select the oils they use in their diet, the choice is probably a matter of taste, or cost.

2. Do you see that green area to the left of the river we are flying over?" said the pilot to the passengers as the commercial 747 jet flew over southern Texas. Continued the pilot, "That's my grandfather's ranch. I often visited there as a kid." Public address systems in commercial planes are now being used by pilots to enliven their passengers' flights. Some pilots are opposed to this practice because they see it as a distraction. Says one seasoned veteran, "Our task... is to fly the plane, not amuse the passengers." But others disagree. Interesting, informative comments can put nervous passengers more at ease and can shorten a long flight. For those pilots interested in making such public-address announcements, there is now a book put together by an ex-pilot which pinpoints more than 1,200 historical and little-known places of interest on a collection of highway maps of the U.S. The maps are overlaid with the flight paths used by commercial pilots. Are you flying between El Paso, Texas, and Las Vegas, New Mexico? If so, then look for the site of the Berringer Crater, where a meteor hit with such force 22,000 years ago that it killed all animal and plant life within 100 miles! Thousands of these maps have been sold, with more frequent fliers than pilots doing the buying. "We are learning...an awful lot about territory we thought we knew!" says one frequent flier who takes her book with her on every flight. Another customer reports, "I bought one for my uncle who hates airplanes, and he now actually enjoys his flights." The book is obviously a success. Is the author correct when he says, "Soon, there will be such books in the pocket of every seat in every commercial flight"?

PART FIVE

Chapters 35, 36, 37 Capitals, Abbreviations, and Numbers

One of Boston's most popular folk heroes is Paul Revere, whose midnight ride was made famous in Longfellow's poem, "Paul Revere's Ride," in *Tales of a Wayside Inn.* Not many tourists realize that many of the places mentioned in that poem are still part of Boston's twentieth-century landscape, along what is commonly called the Freedom Trail. This well-marked trail is actually a two-mile line in red paint, bricks, or steps with twenty-one historic buildings, sites, and monuments documenting Boston's contribution to American history. Two miles isn't long, but it usually takes tourists who are thorough two or three days to see everything on the trail. Most begin at the Boston Common, the oldest public park in the United States. Here colonists used the public land for grazing animals, training the militia, and staging public executions. Immediately next to the Common is America's first public botanical garden, the Boston Public Garden, famous partly because it was the setting for the well-known children's book, *Make Way for Ducklings.* Just across the street, on Beacon Street, is another of Boston's famous sites, the Old State House, completed in 1798. Its cornerstone was laid by Samuel Adams in 1793. One block east on Beacon Street and then one block south on Park Street is the historic Park Street Church on the corner of Tremont. Here the song "America" was first sung on July 4, 1831. The next stop on the Freedom Trail is the Granary Burying Ground, a cemetery on Tremont Street. Some of the greatest U.S. patriots, including Paul Revere, Samuel Adams, John Hancock, James Otis, etc., are buried here. Also along Tremont Street, at the corner of School Street, is King's Chapel, the first Anglican church in Boston. After the revolution, it became the first Unitarian church in the country, and its cemetery contains the graves of many other early notables, including William Dawes, Jr., and Gov. John Winthrop. (It was Governor Winthrop who headed the group that founded the Boston colony.) Also on School Street is the first public school in America, which later became the Boston Public Latin School. Rev. Cotton Mather, Ralph Waldo Emerson, John Hancock, Benjamin Franklin, et al., were alumni of this school. These are only a few of the many sights along the Freedom Trail. Longfellow's poem says, "Hardly a man is now alive who remembers that famous day and year...." True, but the Freedom Trail helps visitors realize many of the events that led to our nation's independence.

Chapter 38 Underlining/Italics

a. U

b. U

c. Q

d. U

e. U

f. Q

g. Q

h. U

i. U

j. Q

k. U

l. Q

m. U

n. Q

o. U

p. U

q. U

r. U

s. Q

t. U

Chapter 39 Spelling

39a. Proofreading

Inventors of gadgets for automobiles haven't always been successful with their inventions. But we can see from some of these inventions that people have been looking for ways to make cars more functional, better looking, and more fun to drive. For example, we now have elegant and sophisticated ways to hear music in our cars, but some of the earlier ways to add music to driving seem a bit odd now. In the 1920s, Daniel Young received a patent for an organ he invented for use in automobiles. He built organ keyboards that could be attached to the back of the front seat so that people riding in the back could play the organ to entertain themselves. This may have been a good idea, except for one thing. The roads of that time, unfortunately, were so bumpy and uneven that the sounds produced by the organ when the car was moving were anything but beautiful. Another terrific idea that didn't make it was Leander Pelton's patent for a car that could be parked by standing it on end. Instead of a back bumper, he built a vertical platform with rollers attached. When parking the Vertical-Park Car, the driver needed to tip the car back onto the platform. Then he could just shove the car into any appropriately sized space. To perform this task, however, was a bit difficult as Pelton never quite explained how the car was to be tipped from horizontal to vertical and back down again. A different problem was that Pelton didn't provide any way to keep gasoline, water, and oil from spilling once the car was up on its parking rollers. But the government gave him a patent; he simply couldn't get anyone to manufacture his Vertical-Park Car. Another invention that never made it was designed by Joseph Grant in 1926—an automobile washing machine. The machine didn't wash cars, but it supposedly washed clothes. Grant's invention consisted of a tub and paddles that bolted to the car's running boards. When the tub was filled with water, soap, and dirty clothes, the bouncing of the car over rough roads provided all the power and agitation necessary to clean a load of dirty clothes. For really dirty loads, an extra twenty miles or so of driving was recommended.

39c. Some Spelling Guidelines

39c. (1) IE/EI

a. believe

b. yield

c. seize

d. height

e. foreign

f. weird

g. field

h. eight

i. deceive

j. financier

k. neither

l. ceiling

m. niece

n. neighbor

o. vein

39c. (2) Doubling Consonants

a. napping

b. footing

c. starred

d. tapping

e. writing

f. referred

g. hopped

h. occurrence

i. beginning

j. benefited

k. shopping

l. omitted

39e. Sound-Alike Words (Homonyms)

a. except

b. accept

c. affected

d. effect

e. here

f. It's

g. it's

h. its

i. passed

j. than

k. then

l. then

m. there, their

n. to, too

o. were

p. who's

q. you're, your

r. advice, buying

s. site

t. quite

u. stationary

v. already

w. all right

x. any one

y. all together

Part Six

Chapter 40 Sexist Language

Responses will vary, but one possible revision is provided here.

In many suburban housing developments built during recent decades there are homeowners associations that enforce housing codes on all the homeowners. The average owners in such a suburb may think that they are free to paint their houses whatever color they like or park any kind of car in their driveways, but that is not the case. Homeowners associations often have a lawyer who spends his or her days enforcing the laws enacted by these associations. The laws see to it that all the members abide by the group's standards of good taste. No plastic flamingos are allowed on the lawn, and all house painters who work in the suburb know that they cannot use certain colors for house trim, such as bright pink or a gaudy yellow, because they have to follow community guidelines. When houses are built in new developments, there are usually restrictive covenants that force buyers to join the association, whether they like it or not. Even when people challenge the laws, they usually lose as the covenants are legally binding. It all starts with the builder because when builders build, they want to make certain that the land value for the community stays high so that they can continue to sell their houses at a good price. The builder often starts off as the chairperson of the homeowners association so that he or she can guide the formation of the rules and regulations. In one wealthy Florida community there are even regulations for local government and civil servants, including dress codes for police officers, taxi drivers, and mail carriers. The only challenges that have gotten through the courts are those that show some regulation discriminates on the basis of race, religion, sex, or other characteristics of the homeowner.

Chapter 41 Unnecessary Words

41a. Conciseness

Responses will vary, but one possible revision is provided here.

Telling time is of great importance to us all, even though the concept of time has changed. Time is important in people's lives because it is the essential measure against which other important measurements are made. For example, we measure our bodies' heart rates and how fast our cars travel in terms of time. We organize our days and nights into segments of time, and from the beginnings of civilization people have counted the passage of time by counting sunrises and sunsets or the movements of the moon and the sun. For centuries, speculation about the nature of time was mostly a philosophical discussion of how people perceive time and experience its passage. But in twentieth-century science, since Albert Einstein, physicists have come to realize that time is a dimension of the physical universe. Time is a measure of motion in space, not some philosophical or theoretical thing that exists in people's minds. Einstein also showed that time is not absolute or unique. People used to think that any event measured in time would be seen to take the same amount of time. That is, two clocks in accurate working order would agree on the time interval between two events. But the discovery that the speed of light appeared the same to every observer, no matter how he or she was moving, led to the theory of relativity. Now time is seen to be relative to the observer who measures it. Each observer can have his or her measure of time as recorded by the clock he or she carries. Clocks carried by different observers do not necessarily have to agree. This is a very different view of time from the older one that time is absolute. However, even with this notion that time is not absolute, we still use time as a means of measurement.

41a-b. Conciseness and Clichés

Responses will vary, but one possible revision is provided here.

Many people today are so devoted to their pets, particularly dogs, that they think they cannot live without them. Whether the pet is a gigantic, wise-looking mastiff or a yapping Chihuahua, all owners consider their pets to be indispensable. Rational pet owners are seldom found. Many dress their pets in expensive clothes they buy on the Internet or in fancy pet stores. They may buy expensive pet beds or send their pets to doggie day care when they go to work. Many would take their pets to work with them if it were allowed. People who don't own pets may consider their pet-owning friends to be crazy because they spend so much money, time, and energy on their pets, but true pet lovers feel calmer with their animal friends and are willing to do whatever seems necessary to keep them happy.

41c. Pretentious Language

2. Responses will vary, but one possible revision is provided here.

As the twenty-first century progresses, American colleges and universities are trying to appeal to diverse students and provide wider access by offering classes in various formats. Students may study in traditional classrooms or complete an entire degree online without ever coming to a campus. Online classes are conducted exclusively through the Internet. If you're considering online education, you should be aware of the personal characteristics necessary for success in this environment. You must be self-disciplined and not inclined to procrastinate, and you must have excellent reading skills. Contrary to what many students think, online classes are not easier, but require much more time and concentrated effort than traditional classes. If you earn a degree online, however, you will have excellent computer skills, as well as knowledge of the subject matter.

Chapter 42 Appropriate Words

42c. Levels of Formality

Responses will vary, but one possible revision is provided here.

Parents concerned about their infants being kidnapped have a new method for identifying the baby. In the past, parents have photographed the child or taken a set of fingerprints to store in computer banks. But these methods are not foolproof. Photographs can be deceiving, especially as children grow up, and fingerprints can be difficult to match. Now scientists concerned about the lack of methods to positively identify a baby have found a promising new approach. Couples can take a sample of their child's DNA material, which contains a large amount of genetic information unique to that child. A repository for storing this information has been set up in New Jersey. The files in this storage facility are then available for matching a sample to a strand of hair, a piece of skin, or a few drops of blood. This DNA "fingerprinting" can positively identify a child at any time in his or her life. The samples in storage can last for the individual's lifetime and many years beyond. Some adults, such as those in high-risk or dangerous occupations, have also become interested in being put on file.

42e. General and Specific Words

Responses will vary, but one possible revision is provided here.

People like driving up for fast food, and now so do their spaniels and collies. A new fast-food industry has begun for drive-in dog food, and the menu is entirely for dogs. These new businesses offer treats to dogs with dog biscuits shaped to resemble the kind of fast food people have. The dog biscuits are made from foods that help keep dogs healthy. The biscuits are flavored differently so that dogs don't get tired of the same thing. Customers love the idea of going to a doggy drive-in after picking up their own fast food. So far the menu has been limited to dog biscuits, but some people will come up with new ideas for better food products for dogs.

42f. Concrete and Abstract Words

Responses will vary, but one possible revision is provided here.

When hikers reach a stream, they often decide to cross where the dirt trail slopes down to the water. But this may not be the best place at which to cross. Water usually moves most swiftly at the narrowest part of the stream. So, hikers should instead look for another spot where the stream widens. Here the current often slows down and may be easier to walk through. When hikers are carrying a backpack, they should loosen the shoulder straps and hip belt before wading in so that they can toss off the pack if they stumble or lose their balance. Some hikers find that if they suddenly hit a depression in the streambed, the weight of the backpack can toss them off balance. Another aid to crossing a stream is a good hiking stick. It can serve as another leg, offering better balance when there are slippery rocks underfoot. It is helpful to remove unneeded sweaters, heavy pants, or boots before crossing a stream with a swift current because the water can drag against waterlogged vests and jeans. The hiker should also take each step slowly and deliberately. The forward foot should be planted firmly before the rear foot is moved. Finally, the careful hiker never hurries across a stream.

42g. Denotation and Connotation

Responses will vary, but possible responses are provided here.

a. famous

b. pushy

c. conservative

d. compulsive

e. sanitary engineer

f. used

g. soil

h. smell

i. cheap

j. lavish

PART SEVEN

Chapter 44 Verbs

1. When people from other countries shop in American supermarkets, they find an amazing supply of items other than grocery items. On one aisle, a shopper will discover, for example, toiletry items such as toothbrushes, toothpaste, deodorant, and shampoo. On another aisle are paper towels, garbage bags, and cleaning supplies. Sometimes the choices are confusing. But the choices do not end with the shopping itself. Even when the shopping has concluded, the shopper may have decisions to make. For example, the person behind the checkout counter may ask if the customer wants the purchases in a plastic bag. Of course, if Americans were to shop in stores and bazaars in other countries, they would find the process just as confusing.

2. In the United States, students hope to learn more than just the subject matter they are studying. They believe that if they study a subject thoroughly, ask questions, and even disagree with certain opinions of writers and scholars, they will become better thinkers as well. Students from other countries may find it difficult at first to practice critical thinking skills in American classrooms because questioning an authority or the written word may be a sign of disrespect in their country. Another problem that students from other countries may encounter in American classrooms involves writing assignments. Whereas American students are comfortable with choosing topics for writing that argue accepted ideas, students from other countries may have difficulty expressing their opinions without apologizing for them because discussing controversial topics may be considered rude in their country. Obviously, students' cultural backgrounds affect many aspects of college life, including how they approach the educational experience.

Chapter 45 Omitted Words

When importing ivory became illegal in the United States in 1989, engineers and material scientists searched for substitutes for natural ivory. For piano makers, the ban on ivory brought on the problem of how to make piano keys after the supply on hand was used up. While conservationists and elephant lovers want people to stop using ivory, many pianists who have tested plastic substitutes are about to abandon hope of having new pianos with satisfactory substitutes. However, there may be a solution. A team of experts has produced a substitute that may satisfy many piano builders and piano players. This new synthetic ivory both resembles natural ivory and has a similar microscopic structure. But even with a substitute for piano keys now available, the market for ivory may not decrease as piano makers have never been major consumers of ivory. The largest consumers of ivory in the world are Asian manufacturers of signature stamps, but even they have given up their use of ivory and are switching to substitutes.

Chapter 46 Repeated Words

a. Students in my class they are looking forward to the upcoming vacation.

b. The book that 1 wanted to read it was not in the library.

c. My teacher found the book that I left it in the classroom.

d. The apartment building where I live there has two swimming pools.

e. The child returned the dog that I had lost it.

f. The pie in the refrigerator it is for tonight's party.

g. The book drop where library books are placed there is full.

h. My aunt called on the new cellular telephone that she had it installed in her car.

Chapter 47 Count and Noncount Nouns

In American supermarkets, new products and produce are constantly being added to the shelves. Children are attracted to new snack foods, and adults are frequently tempted to buy items with information about health benefits. To increase consumer confidence in package labeling, the Food and Drug Administration has announced new guidelines for various claims food manufacturers add to their labels. Products that are advertised as "low fat" have to provide evidence on the label and meet new government standards. For items that appeal to children, the amount of sugar must be clearly indicated. Particularly helpful are the new regulations on serving size because consumers in America have become very conscious of the amount of protein and fat that they eat as well as the number of calories.

Chapter 48 Adjectives and Adverbs

48c-d. A/An/The, Some/Any, Much/Many, Little/Few, Less/Fewer, Enough, No

In northern India there is a conflict between wildlife officials and Gujjar herders of water buffalo. The Indian Government wants to turn the area into a national park, to be called the Fajaji National Park, but for the last ten years, local water buffalo herders refuse to move off the land. The Gujjars keep herding their water buffalo, despite warnings that the animals are eating up too much of the vegetation and that soon there will be few areas that have not been destroyed by the herding. Several decades ago, the Gujjars agreed to migrate every summer to give the forests a chance to grow again, but communities in the areas they migrated to refused to accept the Gujjars because they needed the land for their own grazing, and they didn't have enough land to share. As a result, the Gujjars now stay in the forest throughout the year. Government officials keep on warning of the dangers of erosion in the forests where Gujjar herding has stripped the land. While some environmental groups say that these forest dwellers have as much right to the land as the animals, other groups support the government's attempt to move the Gujjars. The continued grazing by water buffalo risks using up the few food sources of elephants and other animals. Government officials plan to make an offer to the Gujjars to move them to settlements on the edge of the forests and to have them feed their animals in stalls. Park officials want to find a solution soon because they say that both the park and the Gujjars will suffer if the present situation continues.

49. Prepositions

Birmingham in the spring of 1923 was a comparatively new and modern city. Lacking the Old South cultural heritage of the older southern cities such as Atlanta or Nashville, it was characterized by a tremendous energy and pursuit of success that one would expect in an emerging industrial giant. As such, it was somewhat of a melting pot with a constantly increasing number of new residents. Many were from the rural areas of the South, both black and white. There were also significant numbers of immigrants from Italy, Ireland, Germany, and the eastern European countries impoverished by war and other social and political factors. There also were smaller numbers of Asians and others. Inevitably, competition and distrust existed between these various groups, but gradually these were worked out in most cases.

PART EIGHT

Chapter 52 Searching for Information

1. secondary
2. primary
3. secondary
4. secondary
5. primary
6. secondary

PART NINE

Chapter 60 Documenting in Other Styles

1. C
2. E
3. A
4. B
5. D

PART TEN

Chapter 61 Document Design

61a. Placement and Order

The restaurant guide lists ethnic restaurants two ways:
1. By ethnicity

    • Chinese restaurants

    • Italian restaurants

    • Mexican restaurants

2. By neighborhood

61c.

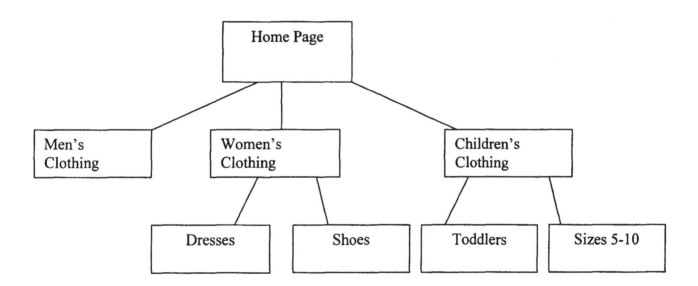

61d. Paper Preparation

1. APA

2. MLA

3. APA

4. APA

5. MLA

6. APA

7. MLA

Chapter 63 Writing about Literature

63d. Conventions in Writing about Literature

1. past tense

2. present tense

3. present tense

4. past tense